Alfred Isaac Bageya holds a college degree in electronics engineering technology with additional training in information technology at York University in Toronto Ontario, Canada. He has worked in various roles in telecommunications systems manufacturing, computer hardware, software support and quality assurance.

He got involved with the Royal Astronomic Society of Canada's Toronto chapter to learn how to build telescopes but fascinated by time measurement problems. He created a simple formula for computing PI (π), which he later transformed into "The General Equation of Time".

He is semi-retired, pursuing his interest in time measurement and other problems that have been largely ignored by mainstream science for years.

To my son, Shannon, and his wife, Lisa, and my granddaughters, Aeisha, Aliyah and Ciara

Alfred Isaac Bageya

THE PI (π) CYCLE SECRET OF THE 360-DAYS YEAR CALENDAR

Time to reset the calendar
back to 360 days a year

AUSTIN MACAULEY PUBLISHERS™

LONDON * CAMBRIDGE * NEW YORK * SHARJAH

Copyright © Alfred Isaac Bageya 2024

The right of Alfred Isaac Bageya to be identified as author of this work has been asserted by the author in accordance with sections 77 and 78 of the Copyright, Designs and Patents Act 1988.

All rights reserved. No part of this publication may be reproduced, stored in a retrieval system, or transmitted in any form or by any means, electronic, mechanical, photocopying, recording, or otherwise, without the prior permission of the publishers.

Any person who commits any unauthorized act in relation to this publication may be liable to criminal prosecution and civil claims for damages.

The story, experiences, and words are the author's alone.

A CIP catalogue record for this title is available from the British Library.

ISBN 9781788487429 (Paperback)
ISBN 9781788786669 (Hardback)
ISBN 9781788787338 (ePub e-book)

www.austinmacauley.com

First Published 2024
Austin Macauley Publishers Ltd®
1 Canada Square
Canary Wharf
London
E14 5AA

Table of Content

Earth to Universe!	9
Introduction	11
Part 1: The Problem with Time	13
Chapter 1: What Creates Time?	15
Chapter 2: The Modern Julian/ Gregorian Calendar	21
Chapter 3: Calendars from Ancient Times	29
Part 2: The Solution for Time and the Calendar	41
Appendix A	57
Appendix B	58
Appendix C	71
Appendix D	77
Appendix E	79
Appendix F	81
Appendix G	83
Appendix H	85
Appendix I	87
References	89

Earth to Universe!

Welcome to a discovery and the process that will transform the way we see, measure and record time with regards to the day, month, year and calendar.

Yes, the way we measure time throughout the year could very well have a whole different premise and foundation than is commonly known—one that is based solely on the rotation of the earth on its axis and its orbital velocity around the sun, eliminating the stars. Ancient civilisations once measured the length of the year correctly—that was 360 days long with 12 months of 30 days each, but when they started gazing at the stars and the moon and tried to integrate them in measuring the length of the year, a perfect system of measurement was broken down and passed on from generation to generation up until now.

This fascinating journey will touch lightly on the history, politics, astronomy, mathematics and religion, and their role in the development of the modern calendar and the never-ending reforms of the latter. This book will introduce two simple mathematical equations that are poised to transform the way we track Earth in its orbit around the sun and the moon around the earth to come up with a perfect 360-day year, consisting of 12 months of 30 days each—the way it was before—with an amazing proof that mathematically, the year cannot exceed 360 days.

Rethinking and re-aligning the year into a new 360-day calendar in future could ultimately have profound results for the world as we know it—and the ultimate benefits of this 'rethinking' may prove to be very profitable, a problem-solver, and quite simply, the year 'put right'.

I will start with a little chat about time and the calendar in general, followed by examining the humble beginnings of the Julian/Gregorian calendar, which is the world calendar now, followed by a brief discussion of time and why it is not the way we think it is. This will not be a challenge to the current scientific theory and philosophical concepts of time but rather, a clarification on where time comes from, giving it a simple mathematical expression. In the second edition of this book, I will provide the methodology for analysis and tracking methods for both the earth and moon, which will establish the much needed mathematical relationships and scientific foundation.

When I am done with my descriptions of time, year and calendar, it will be crystal clear that the earth is the time base generator in the solar system and the universe because of the way we describe time in our daily expressions.

Introduction

The length of the year has been 360 days since the beginning of Earth, time and the calendar

Seasons come, and then come back again, and keep on repeating the same cycle over and over until the earth modifies or destroys them completely. Clouds form and dissipate or dump their excess moisture to the ground and any remaining cloud will hang around for a while or simply drift away, carried by the wind. We are born, and if we are lucky, we grow old and achieve anything within our reach and ability, then die. And like us, all living things, including plants, come to life and die a physical death, never to thrive again with flowers, leaves or fruits that may nourish other living things.

All this happens in space or duration that we measure and call time. Time as we know it is only recorded and used by mankind but no other living thing on the planet cares or records time, but whether it is a plant or animal or insect, as long as it lives, its life is controlled by time.

Ever since human beings developed self-conscience and became aware of their environment, time was already included in the self-conscience package. We have time to eat, to sleep, grow, marry and raise children, hunt, worship, play and die. As humans, this thing we call time now baffles us. Where does it come from? Where is time heading? Scientists, philosophers, writers and politicians are captivated by the subject matter of time. Scientists want to know time in terms of the arrow of time; does it flow in any one direction or is it reversible? What are the dimensions of time? And the questions go on, depending on what domain you are posing those questions from.

Scientists have yet to solve the mystery of time in scientific and mathematical terms, define it and put the dimensions on it.

Then comes the calendar. The calendar, I believe, was designed to record the accumulation of time and schedule and organise time for mankind. Before man came into existence, there was no calendar, or at least, no one needed one;

especially in the early days of mankind when life was about hunting, gathering and creating simple tools needed to guarantee a daily meal then retiring to wait for the next rising of the sun to move on.

Now, almost every aspect of our daily life is governed and directed by the calendar formulated by measuring time from long ago. How time was measured, recorded and accumulated to form the calendar is highly dubious up until now and there are rumours about new calendar reforms on the horizon similar to those made by Caesar and Pope Gregory centuries ago. If this is true, then the problem of measuring time that has dodged mankind for thousands of years still exists; in this book, I will review the problems and suggest solutions for the correct measurement of time.

It is believed that ancient civilisations and cultures had calendars of 360 days in a year and 12 months of 30 days, but supposedly something happened to the solar system after God extended the life of ailing King Hezekiah of Judah by 15 years (Isaiah 38:1-8).

In this book, I will establish evidence based on mathematical calculations and experimental results from 1991-3, that the year is still 360 days long and the months are still 30 days long if measured properly. In a follow-up second edition of the same book, I will introduce the mathematical analysis, which is nothing more than a re-invented circle to create the foundations of the earth and moon, with instruments that can be used by anyone in the world to track the earth and moon in their respective orbits. After 30 years or more, enough evidence will have been gathered around the world to justify the design of a new world calendar consisting of 360-day years of 12 months, each consisting of 30 days, based on the solution I will provide.

Part 1
The Problem with Time

Chapter 1
What Creates Time?

Time is the rotation of the planet through 360 degrees on its axis. Time is created when planet earth rotates through 360 degrees on its axis because a single 360-degree rotation of the earth creates a day, days extend into weeks, weeks extend into months and months extend into a year, thus completing the definition of time. This definition makes time a natural phenomenon created by planet earth in the entire solar system and the universe, making Earth the time base generator in the solar system and universe. This is because every time measurement is an indirect reference to the rotation of the earth on its axis and the number of loops it has made around the sun.

Contrary to modern beliefs, time is not a science and it was not created by science because it preceded science, or stated simply, time does not exist because of science but time created science. This statement, *Time does not exist because of science*, is an absolute truth regardless of the way one interprets it, because attempts to define time scientifically have failed to come up with an acceptable definition of time, including a mathematical expression of time as stated by the General Equation of Time.

What time is not!

Time is not statistical or celestial mechanics, relativity, quantum mechanics or any other branch of science; all branches of science exist because of time. For example, when the first record of time was over 6,000 years ago, there was no science of physics; however, physics and any other science did exist in nature, and as time went on, mankind slowly started uncovering it from nature through observation of the laws of nature and expressing them mathematically.

As mankind started developing science, several philosophers, mathematicians and scientists from all other sciences tried to define time without reference to the earth, but never succeeded.

For example, Aristotle's view of time as "intrinsic and fundamental to the universe" did not mention the planet as the time base generator in the universe because at that time, no one knew that it was the earth that circled the sun and not the other way around, as it was believed before Galileo asserted the fact. Plato's definition was even worse than Aristotle's because the "divine work smith had to impose order" that already existed because of time; not only that, Plato did not think of the fact that he came into existence because of time.

The present time, past and future occur instantaneously

Time has been a mystery and a dilemma when one tries to define the present, future and past. The present time, past and future occur instantaneously, with the future arriving first, followed by the present and the past takes over and continues to drift in the past indefinitely. In the following paragraph, I will analyse an occurrence of a soccer player scoring a goal past the crossbar of the goalpost.

Here is an event; when a soccer player scores a goal, the sequence of events unfolds as follows:

1. The soccer player gets the ball and succeeds in controlling it—the time event is in the infinity because we do not know the player's intentions. They could be one of the following:
    a. The player passes the ball to another player.
    b. He continue
    c. s to control the ball past the defence.
    d. He tries to hit the ball but the defence interferes and the ball bounces back into the field.
    e. The player suffers a heart attack and dies before kicking the ball.
    f. The player kicks the ball towards the goal.
    g. The ball soars in the air past the defence and the goalkeeper.

2. The future starts to unfold:
    a. The ball, in a few milliseconds, could:
        i. Sail over the top of the goalpost crossbar.
        ii. Hit the crossbar and bounce into the field.
        iii. Hit the crossbar and bounce inside the net.
        iv. Sail under the crossbar and hit the net (perfect score).

The perfect scoring event in (iv):

1. Starts when the ball is barely an inch from the crossbar (future arriving).
2. Then the ball sails directly under the crossbar with all coordinates (x, y, z at 0, 0, 0), scoring event.
3. The ball is barely an inch from the crossbar with the coordinates (x, y, z at 0, 0, 1) and continues to Mars along the z-axis—the past kicks in.

The above event is modelled after the final game of the World Cup between Germany and Argentina. In the scoring event, the ball took a three-point flight to go past the Argentinian goalie.

1. The score assist player boots the ball to the relief player (take-off)
2. The ball-scoring player controls the ball with his chest to bring it down
3. Before the ball hits the ground, the relief player boots it past the defence and goalie
4. The ball sails under the crossbar past the one-inch distance of the z-coordinates
5. Score registers at (0, 0, 0)
6. The ball goes past the one-inch mark behind the z-axis—Germany celebrates

In the above scoring event, everyone saw only the present happening but could not see the future because they were too far—even the scoring player didn't see the future arriving because he had to see the ball behind the net, which was in the past.

At the same instant when the future arrived—when the ball was barely an inch approaching the goalpost crossbar—and as soon as the ball was barely an inch away from the goalpost crossbar, within the same instant, the past arrived and continued in the past to the net.

It is this instantaneous occurrence of events in the blink of an eye that has baffled everyone, including scientists and philosophers, who start wondering about the direction of the time flow. From the above analysis, it is clear that time flow is in one direction—the future. The future can be anywhere from 1 microsecond to 1 billion years from now but the past is already out of sight, like the game-winning Germany goal in the World Cup. We have the record of occurrence on video but the time when the event occurred is long gone and will keep drifting further and further in the past as the number of rotations of the planet on its axis and the equivalent linear distance travelled by the earth around the sun increases. The rotations of the earth on its axis and the equivalent linear distance travelled by the earth around the sun are not reversible.

While my definition of time, "the rotation of Earth on its axis", seems very simplistic compared to the volumes and volumes of books that have been written about time by eminent mathematicians, physicists, other related scientists and philosophers, none of them was able to define time. Geniuses like Albert Einstein and Ludwig Boltzmann just mentioned a few developed, highly sophisticated, theoretical concepts that worked seamlessly when implemented, without ever defining time mathematically or in physics.

Without dwelling too much on the past, the rotation of the planet on its axis is what creates time, as we know it. As the soccer player was busy scoring the goal in my simple example, the planet did not blink in its rotation or pause in its orbit around the sun and as humans, we assigned time, date, week, month and year to the event; the planet never cared how we did it but just continued with its business of generating time. In Appendix A of this book, I will define time mathematically, and later in the second edition of this book, will introduce a measurement system or method for time, and surprisingly enough, it will be unidirectional, telling us that time flows in only one direction and is not reversible.

For anyone who will care to agree with me, it will be clear that when an event happens at any instant, it can be defined by a minimum of four mathematical coordinates as they relate to the rotation and orbital movement of Earth linearised. What I am trying to say here is that because the planet is in an almost circular orbit around the sun, it generates equivalent linear distances and angles that never repeat after the occurrence of an event. Using that, one can construct the arrow of time starting from infinity to infinity with all events like the big bang, creation or anything in between. Time shall never be zero as per general equation of time because to have time, mass and distance must exist.

The general equation of time that I developed defines the time it takes the planet to rotate on its axis while it orbits the sun in one direction. Therefore, the rotation of the earth on its axis must have a starting time, starting rotation and a starting distance. From the above information, an arrow of time can be constructed. The above information is summarised below, creating an arrow of time.

To construct the arrow of time from the earth's axial rotations and orbital velocity around the sun, one needs to specify a minimum of four coordinates:

1. The angle starting from 0
2. The numerical value of the rotation starting from 0
3. Start time starting from 0
4. Linear distance starting from 0

Then the arrow of time will be a line drawn through the centre of the earth from infinity and terminating in the centre of the sun—the source of gravity that modulates the planet to generate time—and the initial coordinates of the arrow of time will be 0, 0, 0 and 0.

The first 0 based on my description denotes the angle of the arrow, the second 0 denotes the starting time, the third 0 denotes the starting distance and the last denotes the linear distance covered by the planet after 1, 2, 3…n rotations.

Based on the general equation of time, the time it takes the planet to spin on its axis in seconds (choice of smallest quantification of time) is 87676.8982 seconds. In other words, the planet generates 1^0, with respect to the centre of the sun after it spins or rotates through 360^0, on its axis in 87676.8982 seconds—the length of the day.

So after 1^0, the planet will have coordinates of 1^0, 1 rotation, 87676.8982 seconds and 1.6 million miles or 1^0, 1 rotation, 87676.8982 seconds, 2.6 million kilometres on a linear scale. The above four coordinates give us the natural time, which, according to Roger Bacon (1220-92), is God's time.

By converting the circular distance covered by the planet in its orbit around the sun into linear distance, one can measure any time in any increments and plot the entire constellation along the way. These constellations will repeat every 360 rotations of the planet or 360 days, which is the actual length of the year.

Graphical representation of the length of the year that will run continuously without ever repeating a "number" for trillions of miles, hence the never-ending number π or pi. Only the constellations repeat every 360 rotations to indicate the circular nature of time.

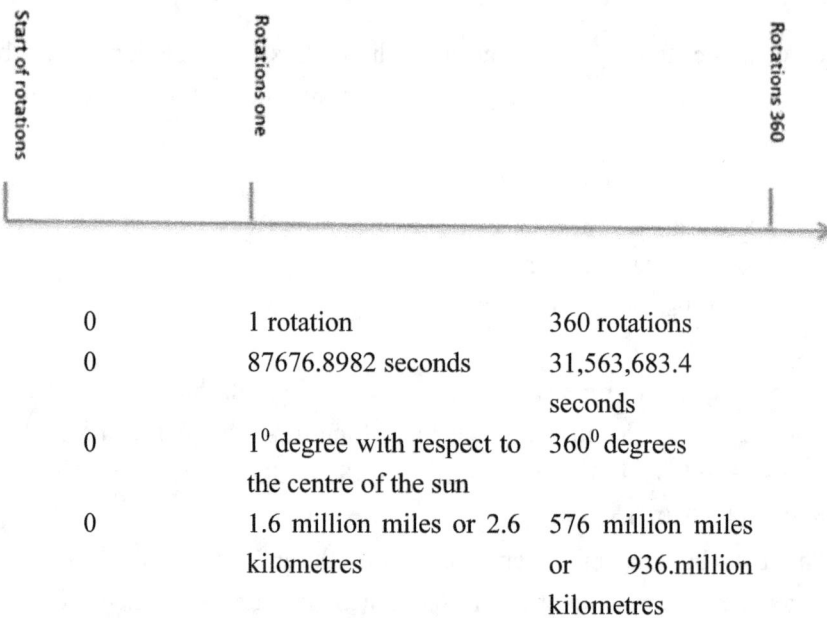

0	1 rotation	360 rotations
0	87676.8982 seconds	31,563,683.4 seconds
0	1^0 degree with respect to the centre of the sun	360^0 degrees
0	1.6 million miles or 2.6 kilometres	576 million miles or 936.million kilometres

Figure 1: Earth's rotations and one-year orbital distance converted to the arrow of time.

Against the above linear arrow of time generated from a single loop of the planet around the sun, constellations can be photographed from the ground or with orbital telescopes/satellites from different distances of the planet, and the telescopes on the equator will be the reference telescopes. The arrow of time in the universe is infinitely incremental; however, changes or shifts in the positions of the constellations can be observed and photographed. By comparing the photographs taken at every 360 rotations, the actual shift in their positions will be detected and there will be no need to reform future calendars because observers will know what is moving in relation to an almost circular orbit of the planet around the sun.

Chapter 2
The Modern Julian/ Gregorian Calendar

The modern calendar has its humble beginnings from two civilisations a few thousand miles apart but share almost the same seasonal characteristics from year to year, although they had no idea that it was the earth that orbited around the sun and not the other way around. Because their experiences with the seasonal changes included fortunes and misfortunes, natural disasters and personal experiences that were imposed on them by nature, ancient time measurement was thought to be a property of local gods. As a result, only the priests who organised their flocks or followers for prayers and offerings to gods were in charge of time measurement because most likely, they were expected to have some form of communication with the gods to determine time for prayers and offerings.

The ancient civilisation with the earliest time record was the Egyptian civilisation, which is thought to have made the first time record in 4241 BCE. It is believed that before the final retreat of the glaciers about 10,000 years ago, the upper part of Africa including northeast Africa was a savanna, lush with rivers and game and inhabited by nomadic hunters and gatherers. When the glaciers retreated—as the earth warmed for unknown reasons, drying up the rivers and killing the fauna and the game 7,000 to 8,000 years ago—the nomadic hunters and gatherers were forced to settle in the Nile valley, which had regular supply of water because of the annual flooding of the river Nile, abandoning the Palaeolithic life of hunting and gathering and adapting to the river cycles. Settlement in the Nile valley by once hunters and gatherers probably created the Egyptian civilisation and also created a deep-set regularity of the Egyptian culture, centred around farming and building settlements, about 7,000 BCE; about 3,000 years later, the Egyptians established what is thought by chronographers to be the first date in human history, calculated to be as early as

4241 BCE. What followed in the next 1,500 years was the unification of a multitude of kingdoms of the Nile into a single political entity and the creation of a complex and homogeneous civilisation with a central authority and religion for 3,000 years until the death of Cleopatra; all this time, the Egyptian civilisation depended on the rhythms of the great river—the Nile.

It is speculated in some books written about the calendar and time measurement that the first Egyptian year was based on the Nilometer, which was a structure for measuring the Nile river's clarity and water level during the annual flood season. As time went on, the Egyptians developed other methods of measuring the year and one them was using the constellations at night supported by the sundial or shadow clocks during the day. Using the shadow clocks, the Egyptians were able to divide the day and night into 12 hours, with the shadow clocks measuring the daytime hours while the successive rising of constellations lasted 10 days and changed with the rising of the sun thereafter. This method yielded 36 constellations lasting 10 days throughout the year and resulted in a 360-day year.

After using the 360-day calendar with 12 months of 30 days each, the Egyptian astronomers tried to equate the phases of the moon with the sun; meaning that when it is full moon, the moon and the sun would be directly opposite to each other with the earth in between, as opposed to the conjunction of the moon when the illuminated side of the moon is directly facing the sun. Under certain conditions, when the moon is close enough to Earth, its shadow will touch the earth, momentarily causing an eclipse. Because this effort did not succeed, the Egyptian astronomers started using the sun as the ultimate measure of the year, thus creating a solar year, which did not sync with the original 360-day year measured earlier using the 36 constellations, which changed every 10 days. The Egyptians erroneously thought that this was an error caused by the moon so they created a mystic story of Thoth the Moon God and added 5 days to the length of the solar year more than 6,000 years ago, creating a year of 12 months of 30 days each. The additional 5 days were tacked to the end of the year to accommodate the mystic Thoth's story. Technically, the 5 days were not part of the 360-day year and this was the beginning of the 5.25 extra days in the year; clearly, it was not scientific or mathematical and was intensely attacked by Roger Bacon (1214 or 1220-92), an English philosopher and Franciscan friar, who asserted that studying nature through scientific experiments, including time measurement, was the only way of arriving at valid conclusions. For Roger

Bacon, knowledge of mathematics was key to unlocking the secrets of the world and saw nothing scientific or mathematical in the belief that the length of the year was 365.25 days.

Although some books on time measurement and the development of the calendar claim that it is unknown how these Neolithic people figured out that the length of the year was 365.25 days, some writers assert that it was the desire by the Egyptian astronomer to sync the flooding of the Nile delta with the rising of Sirius that brought about this change; I believe that this assertion is true. This is because at the time, the ancient astronomers did not know the stars or constellations and that the wandering planets were in a motion of their own as well as the sun. So after the planet looped around the sun covering a distance of more than 500 million miles, it is not possible for the earth-based observer to find the star or planet in the location it was a year ago. Especially with the star Sirius, the Egyptian astronomers found out that they had to wait for an extra 5 days after the arrival of the floods in the Nile delta for Sirius to rise with the sun within the centre of two mountain ridges as an annual interval timing tool for the star.

King Romulus, founder of Rome, and his calendar

By the eighth century BCE, or before, several ancient civilisations had a 360-day calendar. The following civilisations had a 360-day calendar that was later modified to be 365.25:

1. Incas of Peru
2. Mayans of Yucatan
3. Ancient China
4. Ancient India
5. Ancient Babylon
6. Ancient Persia
7. Ancient Assyria
8. Ancient Rome

There is no logical reason why so many civilisations made wrong astronomical observations at the same time and there is no indication as to when Rome actually had the 360-day year. In the following section, I will present the creation of a calendar, which would later be merged with the Egyptian calendar to create the Julian/Gregorian calendar, which is the de-facto calendar for the entire world as we know it today.

The legendary King Romulus, the founder of Rome in 735 BCE, is said to be the creator of the Roman calendar thousands of years after the Egyptian had created and perfected a 360-day calendar, only to modify it later to 365 days, then 365.25 after the Romans led by Julius Caesar conquered Egypt and forced the 365.25 calendar upon the Egyptians.

The first Roman calendar created by King Romulus consisted of only 10 months and 304 days and nobody knows why he set it that way. Romulus' 304-day calendar was not long enough to accommodate all the four seasons experienced in the northern hemisphere and it was short-lived. In 700 BCE, his successor, King Numa, added two more months to make the length of the year to a standard 354 days. Because of the Roman superstition of even numbers, they added an extra day to make 355, and just like that, the Romans had a 355-day year of 12 months without measurement or scientific investigation—just superstitious belief.

The 355-day year was later found to be flawed and months and days had to be intercalated in it to keep it in line with the seasons. Several schemes were devised to correct the calendar, including adding one month every two years, which made the year 366.25 days long; in the end, none of the schemes worked. Then they tried adopting the Greek calendar that inserted one intercalary month every eight years, bringing it closer to 365 days, but the priests did a sloppy job of inserting the intercalary months at the wrong time. The overall solution failed and made the calendric time oscillate back and forth against the solar year.

In the end, the Roman calendar was nothing more than a game between politicians and priests, with each group controlling that part of the calendar that was in their best interests; the bickering went on until 45 BCE when Julius Caesar, then the supreme Pontifex or Pontifex Maximus, decided to take control of the situation.

In 45 BCE, Julius Caesar initiated calendar reforms designed to fix the Roman calendar, which was two months off the solar year at that time. With the help of an Egyptian astronomer, Sosigenes, Julius Caesar reformed the Roman calendar based on reforms ordered by Ptolemy of Egypt in 238 BCE, which made the length of the year equal to 365.25 days with the .25 of a day left to accumulate every three years and becoming a full day in the fourth year to make the year 366 days long—the leap year. Despite the reforms by Ptolemy, the Egyptian priests resisted adopting the 365.25-day calendar and maintained the old 365-day calendar, which had the 5 days extra at the end of the year.

In fixing the Roman calendar to make it similar to the Egyptian calendar, Caesar committed the ultimate error that is haunting the world calendar to date, by sprinkling the 5 days in the months of January, March, May, July and August; from an unknown source, 2 days were added to October and December, leaving the .25 of the day to accumulate every 3 years to create a leap year of 366 days in the fourth year. By sprinkling the 5 days in the five months, Caesar created a virtual planet that orbits the sun in 360 days and our planet, which orbits the planet in 365.25 days and 366 days in the fourth year.

Julius Caesar's fix to the Roman calendar was followed by the most spectacular manipulation of the calendar ever known in history. Caesar initiated it by ordering that two intercalary months consisting of 33 and 34 days be added to 46 BCE between November and December to bring the calendar back into alignment with the March equinox that traditionally occurred on 25 March. Prior to that, an intercalary month had been added in February and when the dust settled down, the year 45 BCE ended up having 455 days and was known as the year of confusion. Caesar's another fix to the Roman calendar was to move the beginning of the year from March to January; by sheer luck, he was right in this because even the random experiment I performed to measure the length of the year agrees with Caesar's decision mathematically.

To enhance King Numa's 700 BCE erroneous fix to the year, Caesar added 10 days to the year to bring it from 355 days to 365 days. Remember that King Numa added 2 months to King Romulus' 10-month year to make the lunar year 354 days long but Roman superstition dictated that the number was an unlucky number so he added one day to make the year 355 days long. Caesar's calendar had 12 months with alternating lengths of 30 and 31 days, except for February, which had 29 days during the 365.25-day year and 30 days for the leap year of 366 days. This was just another icing on the cake of confusion, which, as we will find out later, was ripped apart by Roger Bacon. This is because you can't have 12 unnamed months of alternating lengths of 30 and 31 days long then have an additional month of February that will be 29 and 30 days long—that will be a 13-month year with more than 366 days, regardless of the year.

To honour Caesar after fixing the calendar, the month of Quintilius was changed to Julius—now it goes by the name of July. Caesar's new calendar went into effect on 1 January 45 BCE, with the Romans hoping that they had the most accurate calendar in the world, but that was not true. Soon after his death in 44 BCE, the church priests responsible for the calendar started counting leap years every three years instead of four as decreed by Caesar.

The problem created by counting leap years every three years persisted until 8 BCE when Emperor Augustus fixed the error by ordering three leap years to be skipped.

Then there was another fix to honour Emperor Augustus, the month Sextilis was changed to Augustus containing only 30 days, but the Senate did not stop there, they decided that the month honouring Emperor Augustus should not have fewer days than the month honouring Julius Caesar; so they took one day away from February and added it to August to make it 31 days. After that change, more Roman emperors tried to change names of the months in the year after themselves but the practice did not hold and the names of the months in the year held on to be what they are now.

Since calendars at that time were designed to fix seasons, which are really independent of the stars or the sun that never moves as a centre of the solar system, it was claimed that people knew that Caesar's calendar of 365 days and 6 hours was slow based on the beliefs, but at that time, nobody did anything about it. Nevertheless, we continue using it up to now with the impact of a calendar that adds 5 days and 6 hours to every complete 360-degree loop around the sun.

Roger Bacon (1220-92)

From 44 BCE, after a minor fix to the calendar by Emperor Augustus that saw February losing one day to the month of August, the Julian calendar continued to be the official calendar of the Roman empire and the lands it conquered until 1582. However, Roger Bacon, through his calculations, found that the Julian calendar was 11 minutes longer than the solar year and asserted that the calendar would be off by one day in 125 years and because the error had been going on for centuries, it had accumulated 9 days in Bacon's era, which, if left unchecked, would result in the month of March coming in winter. This is true and it can be verified by the two 360-day cycles I measured back in 1991-3, which show that the earth rejected a total of 7 days—this would have been 6 days but 1992 according to our calendar was a leap year and that accounts for the seven days. Now for the six days left; at least part of the sixth day belongs to the completed year and the other part, if it was measurable, would be the .25 of a day.

These extra days accumulate at a rate of 7 and 6 days every year depending on whether the year is a leap year or what we call a normal year; after 480 years, the planet resets the days and starts all over again. For Roger Bacon, the Julian calendar was unacceptable in all wisdom, which is horror to astronomy and a laughing-stock for mathematics, and wanted Pope Clement IV to fix the calendar.

Roger Bacon went on to assert that natural time was God's time and that time as interpreted by the church could be wrong and mistaken and therefore created three categories of time different from natural time, which belongs to God; the three categories of time included:

1. Time as designated by nature as the passage of years, seasons, months and days.
2. Time as designated by authority when used in civil and ecclesiastic calendars
3. Time as designated by custom when people arbitrarily imposed periods of time, such as months that number 28, 29, 30 or 31 days

Prior to that, Bede had concluded that God's time superseded other authorities. Since Bacon believed that natural time was God's time and time as interpreted by authority such as the church can be mistaken, this implied that whenever an error was found in the calendar, which is time, by authority then the church was responsible for correcting the error.

But on the other hand, Bacon believed that through human endeavours and experiments, errors made by the church—the authority at that time but now is governments and the international astronomic union—could be addressed and possibly fixed.

And like Roger Bacon, that is what I am trying to advocate in my book—to fix the calendar through experimentation and human endeavour. The only problem is that it requires big endeavours, requiring international co-operation in devising new methods for tracking and measuring time because current methods and beliefs are the sources of the 5.25 days that don't exist in the year.

The starting point of the new tracking methods for both the earth and the moon in their respective orbits is to have a ground reference based on longitudes and latitudes for the earth and angular time for the moon because the moon cycle is a subset of the earth cycle consisting of 12 cycles of 30 days each.

For the earth, the reference point will be longitude 0 and this reference point will officially start when it is pointing at the centre of the sun.

At this point, an incremental timing device or clock will be started simultaneously with a cyclic timing device. The incremental device increments indefinitely and the cyclic device resets every 24 hours or more to measure the length of the day based on my calculations.

Chapter 3
Calendars from Ancient Times

According to Dale Wong, "calendars are artificial inventions of man". I agree with Dale Wong's definition of calendars, and I want to extend it to add the most important missing part.

Calendars are failed artificial inventions of man to control nature. And for the last 2,000 years or more, that is exactly what mankind has been trying to do—control celestial objects, including the sun, earth, moon and the stars to agree with what he writes on the papyrus, stone, bronze or iron slabs or millennium's paper calendars. If what he sees in the sky does not agree with what is on the stone slab or paper, he creates a new paper or stone slab to control what is in the sky, or heavens, as astronomers prefer to call it. Below is a list of calendars from ancient cultures dating back as far as 5,000 years, which if compared to the time they have been orbiting the sun is a very small period of time:

- Ancient Egypt
- Ancient Babylon
- Ancient Assyria
- Ancient Rome
- Ancient Persia
- Mayan
- Ancient Mexico
- Ancient China
- Ancient Peru

The idea behind all of the above calendars was to measure time using celestial objects they had no control over. What is interesting about the above cultures is that they all once had calendars that measured time correctly; i.e., 360-

day calendars, but all of a sudden they started believing that the length of the year or the time it takes the earth to orbit the sun was 365 and later they even added on .25 of a day to make 365.25 days in a year.

Only the Bible prophecies kept and maintained the 360-day year for some unknown reasons; however, after reading a few books about the origin and development of various calendars, it appears that the main reasons the ancient calendars started off with 360 days in a year was that the observations were ground-based, especially with regards to the Egyptians.

However, as time went on, keepers of the time started recognising star patterns and an almost regular appearance and disappearance of the moon. So they decided to include the stars and lunar cycles, which are irregular, in their time measurement records.

Unknown to these ancient astronomers, priest timekeepers, political and civic leaders was the fact that the stars are not fixed and the gravitational perturbations of the earth and the moon caused their orbit to oscillate in ways that were difficult to observe with ordinary eyes.

The Egyptian 360-day year—how they created it

Take the Egyptians, for example. They had the correct measurement of the length of year at 360 days using the flooding of the Nile delta for several years, and then noticed that the Dog Star Sirius would rise with the sun at almost the same time the Nile delta was flooded. So the Egyptian astronomers decided to include Sirius's heliacal risings with their 360-day year, only to find that Sirius consistently rose probably 5 days after the flooding of the Nile delta. So to synchronise the heliacal rising of Sirius with the flooding of the Nile delta, they had to add 5 days to the end of the 360-day year and created a mystical story to justify their action. As a matter of fact, the Egyptian priests who controlled the calendar kept a 360-day calendar separate from the 5 days until the Romans conquered Egypt and imposed the 365.25-day calendar, which was reformed with the help of a Greek astronomer from Alexandria—Sosigenes. Without telescopes or cameras, the Egyptians used another ingenious method to record the 360-day year.

Every 10 days, they observed that a new constellation rose with the sun (and on each of the 9 succeeding days, it rose 4 minutes earlier), yielding a set of 36 constellations. With new constellations every 10 days multiplied by 36 different constellations, they were able to ascertain that the length of the year was 360 days.

The day

The day is the starting point of time measurement after hours, minutes, seconds, milliseconds and so on. There are at least two definitions of the day and both are not measurable. The first definition states that a day is the time it takes the sun...to come back to the same location—this is the solar day. It is also referred to as passage of time with reference to the position of the sun in the sky. It is this definition of the day that has created a lot of problems for both ancient and modern astronomy because by definition, it is based on the apparent motion of the sun in the sky.

There is nothing wrong with this definition of the day to the eye, but in reality it is the rotation of the planet on its axis that determines the position of the sun in the sky at any time when the sun is visible to an earth-based observer.

There is another key word in the definition—the passage of time. It is not possible to measure the passage of time because there is no reference point to start from. The original clocks with three arms—the hour, the minute and the seconds—have no reference point. They just move in circles in their enclosures repeatedly.

The position of the sun in the sky can be deceiving as it depends on how fast the planet is rotating on its axis, which depends on its location in the orbit around the sun. The precession of the planet and the area illuminated by the sun at any one time further complicates the phase position of the sun in the sky. The last two factors create an artificial difference between the time measured by the clocks and the position of the sun in the sky because the position of the sun in the sky is observed (not measured) on the sundial while the clocks measure the mean solar time. This is like comparing apple to oranges because the sundial is dynamic and moves independently of the clock. When the clock runs out of battery or breaks down for some other reason, the sundial will not stop and wait for the clock to start. In other words, the sundial runs forever as long as the planet rotates on its axis. This means that any tool designed by mankind to measure time cannot define time but it can measure time because in this case, mankind sets the terms of the operation of his/her tool with respect to the rotation of the planet on its axis. The approach can include atomic clocks or the oscillation of the caesium atom.

Measuring the rotation of the earth against man-made time, like an hour, minutes, seconds or even nanoseconds, is straightforward—just select any reference, including the stars, and wait for the planet to rotate through the reference and record the unit of time selected. This is different from waiting for the earth to rotate through the selected reference, then recording the unit of time selected to measure the rotation of the planet; then you have a problem.

What if there is a noticeable variation in the time it takes for the planet to rotate on its axis? Has the reference moved or the rotation of the earth moved? This is what mankind has been doing for the last 6,000 years or more, and it is not correct. This can be clearly demonstrated by looking at the way the Egyptians added the 5 days to their calendar after measuring the length of the year correctly using the Nilometer, the first instrument they used to measure the length of the year, which they found to be 360 days.

Then they noticed Sirius, the Dog Star, trillions of miles away and decided that it would be a good idea to incorporate it in the measurement of the length of the year. They believed that when the Dog Star was observed rising with the sun, it would mark the end of the year. The results were not that great because it appears that the Dog Star kept on rising with the sun 5 days after the flooding of the Nile delta. Unknown to them was the fact that the Dog Star or celestial objects in the universe has its own itinerary or schedule. By the time the earth completed its own trip around the sun, the Dog Star's position had slightly changed and they needed the extra five days to catch up with it. And this is exactly what happens with our solar year. Whatever the astronomers use as a reference for measuring the solar year—the solar day is on the move constantly—and to compensate for the moving reference, we adjust our clocks; after 400 or more years, we need to adjust our grand clock—the calendar—to bring it back in alignment with the sun, which has never moved an inch since the time astronomers, the church or politicians set the reference.

The result of trying to make man-made time control natural is the differences recorded when measuring basically the same entity, the day. So the general belief is that the solar day is four minutes longer than the sidereal and thus, we end up with two types of years whose mathematics does not quite add up, like having the solar day recorded as four minutes longer than the sidereal day. With this on record, current knowledge about the solar year and the sidereal year is that the solar year is 365 days, 5 hours, 48 minutes and 46 seconds. The sidereal year is measured at 365 days, 6 hours, 9 minutes and 9 seconds, and this is where the mathematics of the day and year measurements fail. If the solar day is 4 minutes longer than the sidereal day, I would expect the extra minutes to be included in calculating the solar year, which would put the length of the solar year to 365 days, 5 hours, 48 minutes and 46 seconds plus 4 extra minutes for every day. I will let the reader do the math as I am probably missing something here because I come up with a solar year of 366.27 days or 366 days, 6 hours, 27 minutes and 23.8 seconds, making the solar year always a leap year with extra hours. Compared with the sidereal year of 365 days, 6 hours, 9 minutes and 9 seconds, there is a difference of 1 day, 0 hours, 18 minutes and 15 seconds. The culprit here is the star we are using to measure the sidereal day—it "appears" to rise 4 minutes earlier.

The sidereal day

Measuring the sidereal day, which generates the sidereal year, is even worse than measuring the solar day. It assumes that somewhere out there is a star (sometimes referred to as a fictitious star) that rises to mark the completion of the 360-degree rotation of the earth on its axis. Below are the two definitions of the solar day and the sidereal day:

Sidereal day	Solar day
The sidereal day is the time it takes for the earth to complete one rotation on its axis with respect to the "fixed" stars. Fixed means that the stars are treated as if they were attached to an imaginary celestial sphere at a large distance from the earth.	*A solar day is the time it takes for the earth to rotate on its axis so that the sun appears in the same position in the sky.*
The sidereal day is 23 hours, 56 minutes, 4.091 seconds, and is shorter than the solar day measured from noon to noon.	*How do we know it is the same position in the sky?*

As you can tell from both definitions, the reference points are floating celestial objects in constant motion and obviously hard to pinpoint.

This clarification about the sidereal day blames the earth's orbital motion, which generates the constellations. Without the constellations, the Egyptians and other ancient and modern astronomers wouldn't be able to estimate the length of the year. Another definition of the same day states that the length of the sidereal day is "the time between two consecutive transits of the First Point of Aries. It represents the time taken by the earth to rotate on its axis relative to the fixed stars, and is almost four minutes shorter than the solar day because of the earth's orbital motion." And we know that stars are not fixed after all.

Summing up everything in the measurement of the day, which is a fundamental measurement of time, we have the following definitions of the global year, not mentioning the religious and other years as defined by various cultures of the world:

1. The solar year: 365 days, 5 hours, 48 minutes and 46 seconds
2. The sidereal year: 365 days, 6 hours, 9 minutes and 9 seconds
3. The Julian/Gregorian year: 365.25 days or 365 days and 6 hours
4. Lunisolar years: collectively at 354 days plus intercalary months and days to catch up with the Julian/Gregorian year

The week

The week just takes on days as dictated by the basic measurement of the day. It is a subdivision of the days in a month and serves to bring about rest of the days within the month or year.

The month

The word "month" and "moon" mean the same thing. The moon is the light of the night sky and orbits the earth in a regular and reliable cycle whose duration in days has been hard to nail down using its waxing and waning pattern times.

The month is the result of the moon orbiting the earth and closing the loop of the orbit. The length of the month depends on observer culture and religion; therefore, a few definitions depending on the culture, the observer, location and many other factors. Like the day, I believe that there should be only one scientific definition of the length of the month that is traceable, measurable and can be mathematically proven. Or improved on in the event of disagreements.

The month was one the first celestial objects of choice for measuring longer periods of time that would accumulate into a year. It was used by King Romulus, the founder of Rome in 753 BCE, who created the Roman calendar with only 10 months, divided into six months of 30 days and four months of 31 days for a total of 304 days. It was King Numa who added another two months to come up with a calendar of 12 months and 354 days but 4 being a bad-luck number, King Numa added one day to make the year 355 days and that was the start of the month music that saw new months with different lengths come and go in the Roman calendar until Julius Caesar took control of the situation in 45 BCE. Today, we have two types of months, the synodic or lunar month and the sidereal month.

The synodic month, or lunar month, is believed to have 29.531 days while the sidereal month is 27.322 days long. Again, the differences between the lengths of the months are blamed on the earth's orbital movement, which is not correct. The reason we see the differences in the lengths of the months, which happen to be the same moon in orbit around the earth, is because we measure the length of the month from the wrong side of the road. Like the earth, the moon's orbit around the earth is almost a circle; therefore, a single loop around the earth must equal exactly 360 degrees. With the lengths of the duration of the moon's orbit around the earth, one cannot archive the 360-degree measurement contained in the moon's loop around the earth. The trick is to measure the length of the month in terms of rotations of the earth on its axis. For every 360 rotations of the earth on its axis, the moon resolves only a fraction of its 360-degree loop around the earth, or 12 degrees. So despite the variations in the moon's velocity and the earth's axial speed, after 30 rotations, the moon cycle or month will be completed because $12 \times 30 = 360$. Stated differently, 30 rotations of Earth on its axis are equal to one 360-degree loop of the moon around the earth and since every 360 rotations of the planet equal one day, then the moon cycle must be 30 days long.

However, the current belief that because of the earth's orbit around the sun, the moon must orbit more than 360° to get from one new moon to the next is probably not correct because the two instances on the month are measured differently. One of them is from first sighting in the western sky to the next sighting in the same western sky (or the duration between successive new moons), while the other is with respect to the sun and stars. As a result, current belief is that the synodic month, or lunar month, is longer than the sidereal month with the sidereal month paged at 27.322 days, while a synodic month is believed to be 29.531 days long.

It is true that if the moon is near apogee at the end of one sidereal month, the moon's orbital velocity around the earth drops, the earth could be in its **perihelion**, which is the point in the orbit of a celestial body where it is nearest to its orbital focus; in this case, the sun. Also, the earth could be in its **aphelion**, which is the point in the orbit where the celestial body is farthest from the sun. In both cases, the axial rotational speed of the earth on its axis is not considered; however, it is a big factor in determining the length of the month because the moon orbits the earth in the same direction of its axial rotation. With the right observational instruments, this factor can be documented and accounted for in the length of the month.

Tracking the moon using the rotation of the earth against the angular movement of the moon eliminates the problems with **lunar apogee**—the moon's farthest point from the earth in its orbit—and **lunar perigee**—the moon's closest point to the earth in its orbit—which are taken care of by the variation in the angles generated by the moon for every 360-degree rotation of the earth on its axis. The sum of the angles generated by the moon every day will always be 360 degrees as long as the moon does not stop or pause in its orbit or the earth does not stop or pause in its rotation on its axis. The influence of the orbital progression of the earth around the sun will be eliminated.

Short or long lunar or synodic month all complicates the accuracy and our ability to predict the length of the month with reasonable certainty. If the average lunar month (new moon to new moon) is about 2.22 days longer than the sidereal month (one complete revolution of the moon relative to the background stars), what months of the year are assigned the 2.2 days? Even if one picks any months at random and adds 2.2 days to their lengths, the results will be inconsistent—a 365.25-day year. For example, 27.322 + 2.2 = 29.522 and 29.531 +2.2 = 31.731. None of these months will be practical to measure and the overall length of the year must be increased to accommodate them.

Overall, the current methods of tracking and recording the length of the month are problematic and not accurate in addition to being incompatible with the solar or sidereal year because none of them will yield days that are equal to the length of the year. Let the brave early adopters try tracking the angular resolution of the moon against the 360-degree rotation of the earth on its axis.

The year

The year is the longest of all time measurement units and it is an accumulation of days from the unit "day", which is the fundamental unit in time measurement. It stands between the seconds, minutes, hours and weeks, months and years. The four types of years in the modern calendar have been mentioned earlier but in this section, I will focus on the "official" year of the

Julian/Gregorian calendar. The modern year is generated by what most books on time measurements have called the most fundamental unit of time measurement in most societies. The day is defined to be the period of the earth's rotation around the sun. The only problem here is that earth has never and will never rotate around the sun—rather, the earth orbits the sun once every 360 rotations (days) on its axis.

I personally believe that a day is a combination of night and day and is equal to one 360-degree rotation of the earth on its axis. The earlier definition of the day used by most books on time measurement creates a number of illusions for lack of a better description or let me call it apparent belief, which puts the earth in the centre of the constellations, making the sun to be moving through the constellations when in fact it's the earth that moves past the constellations as it orbits the sun. This is an ancient belief that should no longer be used because when extended to determine the length of the year, nothing works. This is because the so-called constellations are stars that ancient astronomers and astrologers had created and associated them with human destiny and fortunes. This was not wrong, but to think that the sun travels through constellations is fundamentally wrong and will never give us the correct measurement of the length of the year.

This means that there should be one standard definition of the day based on the rotation of the earth on its axis, regardless of the position of the sun in the sky. In this book, the word "day" is reserved for one complete rotation of the earth on its axis, which is the same as 360-degree rotation. There is also a need to a single point of reference to be regarded as the starting point of the rotation of the planet on its axis and because the rotation is a perfect circle, then it does not matter at what point in space we start this rotation; the earth was, is and will always be rotating on its axis and at the same time orbiting the sun.

Once the starting rotation of the planet is determined by agreement from all nations, then the starting point will have three parameters: the starting angle = 0, rotation = 0, at mile/kilometre = 0 and starting time = 0. No calendar date shall be assigned because they are not reliable and this will be the starting point of a new time measurement system, which can be implemented after 25-30 years to create a new calendar.

Furthermore, the starting point should be longitude 0 as close to London, UK, as possible. This is where a mental pause of the planet's rotation on its axis is pinpointed, then rotate and advance the planet in its orbit around the sun to initiate the four coordinates of time measurement: angle, rotation, mile/kilometre and time at (0,0,0,0). After one full rotation of the planet on its axis, the new coordinates of the planet will be (360 degrees, 1, 1.6/2.6, 24); each country will have the option of expressing the middle parameter in miles or kilometres. This tracking method and recording of the planet's location will continue until coordinates (129,600 degrees, 360 rotations/days, 5.76 x 10^8 miles or 9.36 x 10^8 kilometres, 8.6x10^4 hours), after which a yearly counter is incremented by 1. After 30 years, the coordinates of the planet shall be (38,888,000 degrees, 1.08 x 10^4 rotations, 1.728 x 10^{10} miles or 2.808 x 10^{10} kilometres, 2.58 x 10^6 hours).

The reason for this approach is to create a linear arrow of time in which every event has its own linear coordinates that never repeat and to end the debate on whether time is reversible. This method will give generations and generations of mankind a linear scale on which all events, including birthdays, are calibrated with reference to $t = 0$ or since this method of time measurement was implemented using this method. This approach will also answer questions like: how far away the earth would be if we started measuring time this way 5,000 years ago? Note that the number of days in a year will remain constant at 360 days forever because of the way the two variables, R and M, track each other. Any change in one variable results in the change of the other variable in order to keep the number of degrees traversed by the earth constant at 360^0 and the value of pi constant in the dynamic formula for pi.

The backdrop of constellations will be plotted on the linear scale and their variable visibility will oscillate back and forth to resonate with the precessions of the equinoxes; there will be no need to reform the calendar in order to align it with the sun or stars because we know that they are moving and we are moving. With regards to the principle of natural day, this method of tracking the planet does not change the principle.

The natural day will always be a segment of a continuum and can begin anytime. Different cultures and religions will be able to continue with unanchored natural days distinguished from the civil day in a strict sense, which will be the natural day as reckoned from a particular point determined by the law or custom. In Western culture and ancient Rome and China, the natural day starts at midnight; in the Jewish, Muslim, ancient Greek and Babylonian, it starts at sunset; for the Egyptians, it starts at sunrise. The time after midnight remains tomorrow, which by default is considered to be "morning". Astronomers will still be able to measure their day from noon to noon, it will also be the beginning of the nautical day, which will allow all observations relating to a single night to fall on the same date.

As an example, a police officer reporting an occurrence of a crime during the hours after the regulated end of the day but before the regulated start of the day will state that the crime occurred at night even if he gets at the crime scene one second before the start of the day. The same will happen if he/she arrives at the crime scene one second before the end of the day. (I have used crime as an example because it is the fastest recreational activity in the world and to show how mankind is fast-forwarding itself in obviation).

In the next section, I will explain how this method of defining a civic day will fit into the strict method of defining the length of the day based solely on the 360 rotations of the earth on its axis as it will come with a mile or kilometre post in space.

The cumulative kilometres or miles represent the equivalent linear distance covered by the planet at any instant of time, thus representing the arrow of time that has been a subject of investigation by physical scientists and mathematicians for a long time; they could use "bold insight and intuition which may lead to new concepts barely conceived today" (Peter Coveney and Roger Highfield, The Arrow of Time: A voyage through science to solve time's greatest mystery, P-293).

I call it the arrow of time because time in the solar system is cyclic, while time in the universe is a straight line. Along this straight line, events occur at specific non-repeating coordinates as the universe expands or contracts; that way, time in the solar system and time in the universe is merged into a single measurement. Although scientists and physicists have been asking whether time is reversible, the arrow of time constructed this way dictates that time is not reversible because once an event happens at any location on the arrow of time, time continues to progress in a positive direction from $t = 0$ to infinity while the time to the left of $t = 0$ before the event occurred starts from infinity and no one knows what happened; a little bit of chronology and carbon dating will estimate the possible time when an event happened.

This phenomenon is clearly visible in our immediate surroundings within the universe because all objects visible to us, including the galaxies and stars, are always in motion—moving away from the sun or the sun following them or moving along with the sun or following the sun from behind. So the idea of measuring the length of the sidereal year with respect to a fixed star is definitely incorrect because the stars are not fixed in space. Whether a star is in front of the solar system or following or beside, anytime an earth-bound observer tries to fix its position with respect to the earth, he/she will have a constant error, which observers of the heavens see as 5.25…days. This error is a combination of three factors: the distance between the star and the observer (time dilation at work here), the motion of the star makes its exact location change after the 360 days it takes Earth to circle the sun, the observer's angle of sight is extremely large and therefore it takes a finite amount of time for the star to appear in alignment with the observer or the sun.

This was the case when the Egyptian astronomers tried to sync the appearance of Sirius with the flooding of the Nile. Because Sirius appeared to rise with the sun five days late, they decided to add five days at the end of the year. This was an error but since the astronomers of that time had very limited knowledge of the universe and the solar system, this error was expected but we shouldn't continue making the same error because we know better.

And finally, the idea of measuring the progress of the sun through heavens as a way of measuring the length of the year should be completely abandoned because it is confusing and it doesn't make sense to model the length of the year in two different ways using the same objects—the sun and the earth. Measuring the length of the year using the sun as the centre of reference makes a lot of sense because that is how the solar system is modelled. With this model, it is possible to envision the planet orbiting the sun as modelled in most books of astronomy minus the stars.

Part 2
The Solution for
Time and the Calendar

For thousands of years, it has been assumed, or we have been told, that the length of a year averaged from different types of years (sidereal year, solar or tropical year, anomalistic year, vague year, etc.) is 365.25 days long. This is not true and as a matter of fact, the following days of 31 January, 31 March, 31 May, 31 July and 31 August do not exist. They are just days on the calendar's paper or paper days.

On the other hand, 31 October and 31 December possibly belong to the month of February. So this means that the 5.25 in the year exist on the calendar only and are excess baggage, costing the world economy billions, if not trillions, of dollars in payments for utility bills like water, electricity, natural gas and in automotive fuels like diesel, petrol (gas) for fleet operations and aircraft fuels.

The following example shows typical over-payments for electricity alone per year for the mentioned business operations in North America:

- A software company data centre spending $3.9 billion/year overpays $50 million/year
- An automotive parts manufacturing company spending $16.8 million/year overpays $241,478.44/year
- A medium-sized 200-unit apartment building spending $400,000/year overpays $5749.49/year
- An average home spending $4,800/year overpays $69/year

The economic cost of five and a quarter (5.25) days to everyone

Services for water, electricity and natural gas are billed on a monthly basis, meaning that the utility company charges customers for water, natural gas and electricity 365.25 days annually, including the non-existent 5.25 days.

Let it be known that the 5.25 days exist only on the calendar paper they are written on; this means that every customer for the utility companies ranging from governments to companies, corporations and individual home-owners pay for water, natural gas and electricity for paper days since the 5.25 days don't exist in reality. This is because the calendar dictates the following days for each of the months listed below:

1. 31 January
2. 31 March
3. 31 May
4. 31 July
5. 31 August

In reality, all the 31st days of the above listed months do not exist because, after calculating the length of the year and measuring it, then calculating the days in the month and accounting for the missing hours or days and taking into consideration (and observing) the interaction of the motions of the moon and earth throughout the year (in addition to the two-year experiment and observations performed by the author), on a global scale it can be proved that the length of the year is 360 days and the length of the months is 30 days.

However, natural gas/electricity customers still have to pay for the extra day in the following months because February is always short of one day in a leap year—366 days—and two days in a normal year—365.25 days—despite the fact that the .25 days have never been measured; on the other hand, 31 October and 31 December seem to be floating days because for three years, customers pay for 28 February and in a leap year, they pay for 28 and 29 February.

Basic overpayment examples

1. **Automotive manufacturing outfit**
    An automotive parts manufacturing outfit that spends $1.4 million a month on energy bills (natural gas/electricity). The facilities manager of this automotive parts manufacturing company pleaded with employees to help the company save on energy costs by turning off all lights, especially on the weekends (a drop in the bucket compared to what the machines and processes consume during working hours), except the electrical/gas heaters that keep the building warm in winter and cool in the summer because the manufacturing process is sensitive to extreme temperature swings.

If $1.4 million is multiplied by 12 and divided by 365.25 days of the year, the result shows that the company is paying $45,995.89 per day for energy.

Then multiply $45,995.89 by 5.25 = $241,478.42, the amount of money that the company is paying to the utility companies every year for energy used in the non-existent 5.25 days.

This company can save up to $241,478.42 a year by re-negotiating to pay the utility companies on a per day basis for the energy it uses (natural gas/electricity) and stop paying for the 5.25 days that don't exist.

2. **Telecommunications companies**

Telecommunications companies consume large amounts of natural gas/electricity in their switching centres, which are now doubling as data centres to consume even more power. Telecommunications companies should be pioneers in negotiating with natural gas and electrical utilities companies to change the billing systems to "charge per day" for 360 days a year because the power bills for telecommunications are potentially very high.

It is possible for a telecommunication company to have up to 500 (someone can correct me if I am wrong for the number) switching centres distributed across a country like the US and mirrored with each other to guard against a power failure or fire, and each of them may consume an average of $25,000 a month in natural gas/electrical bills.

[($25,000 x 500 x 12)/365.25] = $410,677.62 per day.

$410,677.62 x 5.25 = $2,156,057.50 in potential savings per year if the company is billed on per day basis for 360 days a year.

3. **A software manufacturing company with a data centre**

There are a number of software companies out there that operate large data centres. One software company spends well over $3.5 billion on energy for powering data centre servers and switches only; energy consumed by the cooling systems for the servers and the building housing them is not included.

This works out to be $9,582,477.75 per day. If that number is multiplied by 5.25 days, it works out to $50,308,000 in savings if it is billed on per day basis to exclude the 5.25 days. (It is estimated that a single data centre server and its cooling unit can cost up to $800 a month in energy consumption; this cost will be higher if the server is running 4 to 8 virtual machines and feeding multiple virtual networks.)

If at least 1,000 companies/corporations that spend more than $3.5 billion per year on energy come together with a will to save $50 million per year on their energy for 10 years, a significant amount can be raised to deal with global poverty.

But before this can be done, there is a need to agree—through discussions, experiments and measurements—with the fact that the length of the year is 360, not 365.25 days. The basic mathematical theory and tools to do the job are available but need development.

4. Real estate industry

The real estate market can be divided into four main markets:

1. Commercial
2. Residential
3. Industrial
4. Vacant land

In commercial, residential and industrial real estate, there is always some kind of building(s) constructed on a piece of land. These buildings can be owner-occupied property, large industrial buildings like warehouses or manufacturing plants, mixed residential and commercial buildings.

Regardless of the type of building construction or usage, the main reason for their existence is because someone invested in one type or multiple types of building to earn income from the buildings; however, common to all these buildings is energy (electricity or natural gas) and water usage. The value of most commercial, residential and industrial properties is almost always dependent on current or projected cash flow and in turn, this cash flow is influenced by the cost of maintaining the building(s) and supply and demand.

If you own an investment property, regardless of the size and location anywhere in the world, regardless of whether you are a large real estate investment company or small holder, you will benefit tremendously from understanding the need to pay for the utilities such as electricity, natural gas and water on a per day basis so that you only pay for 360 days of the year for those utilities. This is because the value of an investment property is largely influenced by the income it produces and any step taken by the investor to decrease the expenses and increase the income from an investment property will have a direct impact of increasing the value of the property immediately.

The traditional methods of increasing the income a property generates every month include raising rents, reducing expenses and making use of rental equipment wherever possible; in addition to those methods, paying for utilities on a per day basis for 360 days a year reduces the expenses and increases the income a property generates right away with a bonus of increasing the value of the property as well.

I will demonstrate the tremendous impact of the suggested pay per day method on the value of property, the cash flow and expenses using a hypothetical 5-unit apartment building valued at $18,000,000 with a cap rate of 6% when the investor converts to a pay per day method.

The assumptions here are that:

1. Water costs: $8,800 per year
2. Electricity costs: $280,000 per year
3. Natural gas costs: $98,060 per year
4. Total expenses on energy and water: $466,000 per year

Paying $466,000 per year for the above utilities translates into paying $466,000/365.25 = $1,275.84 per day.

Now if we multiply $1,275.84 × 5.25 = $6,698.15 extra money paid for the utilities per year.

If the investor pays for utilities on a per day basis for 360 days, it means they pay $6,698.15 less in utilities per year, translating into $6,698.15 increase in the income from the property per year or $6,698.15/12 = 558.18 extra cash in the bank per month.

The property value goes up by 6,698.15/0.06 = 111,635.83 or 18,000,000 + 111,635 = $18,111,635.

For commercial large-scale investment properties, utilities bills run into millions per service (electricity, natural gas and water), resulting in hundreds of thousands of dollars in extra income per year and property value running in millions, if not hundreds of millions, depending on the number of properties a corporation might have.

Before implementing pay per day system anywhere in the world, the issue of year length and the economic benefits should be reviewed to decide whether the length of the year is 360 or 365.25 days and what to do with the surplus cash on energy budgets of every individual home/property owner, companies/corporations and above all, governments and their departments/agencies. The basic mathematical theory and instruments to do the job are available; however, the instruments need development.

The 5.25 non-existent days don't seem to be such a big deal to anyone until the cost of paying for certain goods and services for days that don't exist is taken into consideration.

Paying for services such as water, natural gas and electricity on non-existent days costs everyone, including governments, individual home owners, companies and corporations, billions of dollars annually and possibly trillions of dollars if the cost is added up globally.

The point here is that payment for water, natural gas and electricity in most countries should be changed to pay per day for 360 days a year.

If a change like that is implemented globally, it will help consumers slash billions of dollars, if not trillions of dollars, from their energy budgets. And if the money saved is put into a special fund, it can be used to tackle major social and economic issues worldwide that no one dares to touch because of the cost, cultural beliefs and political norms/abnormalities.

Other benefits of a 360-day year

A 360-day year in future (25 to 30 years from now) will save corporations and governments around the world billions of dollars in overhead costs spent on utilities.

For those salaried folks who are paid on a monthly basis at work, if the base salary of $30,000 a year is calculated on the basis of daily earnings in a 365.25-day year, they make $82.13 per day.

In a 360-day year at the same salary, they make $83.33 per day. That translates into an extra 428.40 per year or instant pay raise.

How and when it was found that the year is only 360 days long

In 1991, I designed a simple experiment using a simple prototype instrument to track the planet one rotation a time in its orbit around the sun from December

1991 to December 1993, and found out that Earth rotated 360 times on its axis before it completed one loop around the sun.

And for the moon, I used another prototype instrument to track the moon in its orbit around and found out that it lost an average of 48 minutes against Earth for every single 360-degree rotation of the planet on its axis. Using simple linear interpolation on the single piece of data I captured for the moon, and assuming a perfect circle, I found out that the earth rotates 30 times before the moon makes a single loop in its orbit around it. I concluded that the moon cycle or the length of the month must be 30 days and the earth cycle or the length of the year must be 360 days.

In order for this to be a qualified scientific experiment, more tracking locations distributed randomly or by design on the entire planet earth are needed and the prototype instrument used to track the planet in its orbit around the sun one rotation at a time needs to be developed so that more people and/or organisations get involved in this new tracking methodology to evaluate the consistence of the collected data from around the world to see if it can be used to predict the earth cycle.

The moon-tracking instrument that tracks the moon in its orbit around the earth also needs development for the same reasons.

Data gathered around the world through an experiment like this can be used to design a more robust and stable world calendar in future (25 to 30 years from now), because the current Julian/Gregorian calendar tends to fail every 400 years or so. The frequent failure of the current calendar has forced unwanted calendar Band-Aid reforms that never fix the problem but provide temporary patching through adding days to the months in the calendar, but mostly by deletion of days from the calendar.

Fixing a failed calendar like this can be done because there has been no way of linking the days (360-degree rotation of the earth on its axis), the weeks and months in a year through a single mathematical equation that enables accurate calculation and measurement of the days and months and eventually the year. Limited information on the origin, design and continued development and reforms of the Julian calendar, which later became the Gregorian, and difficulties encountered in the past 2,000+ years will be presented.

Research for the source of 5.25 days and how they got into the modern calendar

After researching a few books about the design and development of the modern calendar from ancient times to today punctuated by reforms, I found out that the 5.25 days do not exist in the time it takes Earth to orbit the sun, but were inserted into the year by ancient astronomers after they started using the stars and constellations (and we are still using stars and constellations) to determine the start and end of the earth cycle or the year in its orbit around the sun.

Using the stars to measure the length of the year is not a bad idea, but they are not fixed as presumed since the start of time measurement on earth by mankind because all objects in space, regardless of whether they are galaxies, stars, planets, comets or simple chunks of rocks, must be in constant motion to maintain dynamic equilibrium and avoid collisions with other objects in motion. By looking at the Milky Way galaxy, it is clear beyond reasonable doubt that billions of stars (suns), including our solar system, are all in motion towards the galactic centre.

Another interesting observation I have made is that vernal equinoxes are a function of the area illuminated by the sun on a perceptually rigid and inclined planet earth and as a result, they make the sun appear to move back and forth over the maximum and minimum inclination of the planet as observed from the northern hemisphere where almost all activities related to the design and development of the modern calendar originated.

A 360-day year has no impact on religious, political, ethnic/tribal or any beliefs out there as far as time measurement is concerned and will not impact other living things including plants, water- and land-based animals and creatures because the 5.25 days were not part of the earth cycle and were first used in the ancient Egyptian astronomers' calendar, when they inserted them at the end of the 360-day year for special reasons not related to any religion or civic activities, according to information available on ancient Egypt and its history.

The 5.25 days in the Western calendar were introduced by Julius Caesar in 45 BCE during his calendar reforms with help from Egyptian astronomers. Caesar added 5.25 days to the Roman calendar and sprinkled them in the five months of January, March, May, July and August, then added the .25 days to the end of the year to create a leap that doesn't exist either.

The information in this book is a pure mathematical/scientific endeavour to try and convince the readers that it is about time we used a mathematical/scientific approach to determine the earth cycle and hence the length of the year, starting with a measurable length of the day, then the rest of other time measurement units will fall into place naturally because the day is the most fundamental unit on which all time measurements are based or should be based.

In tracking the earth, I created two systems—the Sun-Earth System and the Earth-Moon System, which were analysed independent of each other with full knowledge that the days in a month or moon cycle around the earth are a subset of the days generated by the earth as it spins on its axis to create daytime and night-time. The Sun-Earth System includes records of results of tracking the planet earth in its orbit around the sun for two years.

The planet earth is the Time Base Generator for the solar system and universe at large, so re-measuring the earth's cycle and setting it to a new time cycle for its orbital period may have a non-consequential small impact on the cycles of the planets within the solar system and no impact at all to the time in the universe because of its size. The issues listed below will be the core subject matter of this website:

1. General information on why the 5.25 days in the calendar (Gregorian calendar) don't exist and why they were arbitrarily tacked at the end of the old Egyptian 360-day year by the Egyptian Moon God Thoth as well as a brief description of why the 5 days were added to the end of the 360-day Egyptian year that was used before the Egyptian Moon God Thoth implemented the change.
2. On the SUN-EARTH page, a simple formula for calculating π was introduced; then using simple transformation of formulas, the general equation of time was developed and it is the general equation of time that mathematically fixed the length of the year to 360 days before it was measured practically by tracking the planet earth in its orbit around the sun.

The general equation of time calculates the time it takes the planet earth to resolve 1° with respect to the centre of the sun as it spins on its axis while orbiting the sun at the same time—the planet earth executes the two motions simultaneously. The general equation of time uses the sun as the reference centre for the earth to generate time. Because it is the time it takes the planet to spin 360° on its axis, this is the time that regulates human civic activities, the time used in the design and creation of ancient and modern calendars for centuries; finally, it is the time that regulates other natural activities on earth, including the growth of living things from inception, germination of seeds or growth of roots (for the plants and trees), weather system, volcanoes, seasons, etc.

As it is the time it takes the earth to resolve 1° with respect to the centre of the sun as mentioned earlier, it is also the time it takes the planet to spin once through 360° on its axis. Calculations show that it equals to 24 hours, 21 minutes and 17 seconds or 1 day. This compares well to the length of the sidereal day, which is 23 hours, 56 minutes and 4 seconds, despite a difference of 25 minutes and 12 seconds in favour of the calculated day as opposed the sidereal day, which is obtained by observations of the rising sun.

On the same page, information on how to calculate the length of the year using the general equation of time is provided and it is an extension of the calculation of the length of the day in the form of the time it takes the earth to resolve 1° with respect to the centre of the sun while spinning once through 360° on its axis. It is simply the time it takes the planet to spin on its axis multiplied by 360° in a circle, and the result is the length of the year, which equals to 360 days.

3. A record of two years of observational data confirming that the planet earth takes 360 days to orbit the sun and not 365.25 days as generally believed without scientific or mathematical proof. This is attributed to the fact that before Galileo Galilei first used the telescope for the purpose of astronomy in 1609, there was a general belief that the earth was the centre of the universe and issues related to the calendar and time measurement were handled by the church. As a result, the 365.25 days in a year is neither a religious, scientific or mathematical belief because the idea of tacking 5 days to the end of the 360-day year was to solve a domestic issue in Egypt only, not the world.

Earth-moon system generates the month and fixes it to 30 days

The time it takes for the moon to orbit the earth is derived from the general equation of time by calculating the time it takes for the moon to generate 1° with respect to the centre of the earth then multiplying that time by 360° in a circle to obtain the moon cycle or the time it takes for the moon to orbit the earth. Using the earth as a centre of reference, the time it takes the moon to resolve 1° with respect to the centre of the earth was found to be 1.88444808 hours.

Then the 1.88444808 hours were multiplied by 360° in a circle to yield 678.4130962 hours for the moon cycle around the earth, or a month. The resulting 678.4130962 hours were divided by the 24.354693947 hours or (24 hours, 21 minutes and 17 seconds—the calculated length of the day), [678.4130962/24.354693947 = 27.855053758 days] for the moon cycle or month. This moon cycle around the earth was found to be short 25.739355 days for the 12 months or moon cycles in a 360-day year or 2.144946242 days missing from each month. [(360 - (27.855053758×12)) = 360 - 334.260645096 = 25.739355] missing days in the year or 2.144946242 missing days per month. If the 2.144946242 missing days per month are added to the 27.855053758 calculated moon cycle or month, the result is 30 days in a month—30 × 12 = 360 days in a year.

Out of the 25.739355 missing days in a 360-day year, only 12 days could be accounted for; the remaining 13.739355 days in a year are still missing or 1.14494625 days from each month of 30 days in the 360-day year.

The missing 12 days or 1 day a month can detected by naked-eye observation

An earth-based observer observes the moon passing through a point in space and makes note of the point in space where the moon is observed. After the planet rotates through 360°, the earth-based observer checks the point in space where he/she observed the moon and finds that the moon is not there. This is because of the fact that the moon orbits the earth in the same direction as the earth's rotation. As a result, after a 360° rotation of the earth on its axis, the moon arrives at the point it was observed a day before 48 minutes late. Assuming that the planet spins on its axis 30 times before the moon completes 1 loop in its orbit around the earth and it loses 48 minutes every rotation, the moon loses a total of 30 rotations x 48 minutes = 1440 minutes, which is equal to 1 day missing based on a day of 24 hours exactly.

These results are shown in a dynamic tracking table on the moon page based on a perfect circle.

So in 12 months of the year, we have 12 missing days. If the 12 missing days due to the moon being late 48 minutes every 360° rotation of the planet on its axis are added to the 334.260645096 or [(27.855053758×12) + 12] = 346.260645096 days in the year.

Now we still have 13.739355 days in a year missing or 1.14494625 days from each month of 30 days in the 360-day year as mentioned earlier because 346.260645096 + 13.739355 = 360 days in a year.

The 1.14494625 days missing in each month can be found by designing an experiment that tracks the moon on a daily basis for at least two years or by adjusting the distance between the moon and the earth repeatedly until the value of 360° x t of the moon = 30 days.

The above information combined with the issues and problems related to the development of the modern calendar indicates that there is room for improvement by way of introducing new and innovative approaches to time measurement that could be more robust, probably more reliable and simpler. The improvement needs development of new methods and introducing two new instruments that have never been used before in astronomy to dynamically track the earth and moon in their respective orbits:

1. One instrument will dynamically track the earth in its orbit around the sun daily; therefore eliminating the current method of tracking the earth and sun through the star constellations and the Zodiac, but will have no impact on the way astronomers and astrologers use the Zodiac or astrological beliefs. The earth tracker will be home- and office-friendly.
2. And another instrument will do the same for the moon, tracking it in its orbit around the earth, but differently from the earth-tracking instrument because the moon-tracking instrument will be bulky compared to the instrument for dynamic tracking of the earth.

Combining observational data from the two instruments will prove that the earth cycle or length of the year is 360 days only and that the moon cycle or number of days in a month is 30 days after correcting previously unknown issues. Once the instruments are developed and deployed around the globe, data will have to be collected over a period of 25 years or more (The earth will need to log 9,000 observation days or 25 × 360 = 9,000 to prove that the earth cycle or year consists of 360 days because the parameter used for the observations occurs only once every 360 days.) in order for observers to statistically prove that the data is correct and reliable.

It is possible to make some needed changes right away (like water, natural gas and electricity billing can be changed to per day basis for 360 days) for economic reasons, including alleviating economic hardships imposed on governments (mostly in the northern hemisphere), corporations, companies or businesses and individuals who spend more than $1,000,000 a year on utility bills. (Please read the extra information regarding the economic cost of the 5.25 days in a year.)

The solution for creating the future calendar

The modern solution: First and foremost, the author and David Ewing Duncan acknowledge that the problem exists even today despite the modern instruments being used.

The first assumption to make when creating a solution for this problem of time measurement is to accept the fact that the orbital motion of the moon around the earth is independent of the orbital motion of the earth around the sun and what has been missing is a mathematical link between the two celestial objects we use to measure time. The general equation of time establishes this link and it applies to both the earth and the moon despite the fact that each celestial object orbits a different centre of gravity.

However, the general equation of time cannot apply to other planets because there can only be one time base generator in the solar system and that is Earth. This is because time is measured by mankind and all references to time measurements in years refer to the number of times Earth (not any other planet in the solar system) has orbited the sun.

Based on "simple" mathematical analysis, the earth generates a day when it makes one complete rotation of 360 degrees on its axis and in general:

1. One day starts the time measurement
2. The days accumulate into a week
3. Four weeks accumulate into a month
4. Twelve months accumulate into a year

All of the above time measurement units accumulate without any reference to the sun whose job is to modulate the earth in its orbit around it and the earth's axial speed.

The day

The day will be defined as one 360-degree rotation of the earth on its axis measured in:

1. Hours
2. Minutes
3. Seconds
4. Milliseconds
5. Microseconds
6. Nanoseconds

The month

The moon on the other hand independently slices the 360 days generated by the earth into 30 days a piece to complete one loop in its orbit around the earth—the month. Mathematically explained, the number of days in the month is a subset of the number of days in a year and because the earth rotates 30 times on its axis before the moon completes one loop around it, then the moon must make 12 trips around the earth before the earth completes one loop around the sun.

The week

As for the week, it needs 7.5 days in order to fit evenly into the 30 days of the moon; however, the .5 of the day cannot be measured because the rotations of the planet on its axis to generate the days are discrete—they come in 1, 2, 3…30 with no fractions.

Therefore, to accommodate the fractional day of the week, it must be deferred for later measurement. This will result in deferring two days of the month in a four-week month of 30 days, which will later be added to the month of December; by the end of November, December will have the 30 days needed to have a year of 360 days because at the beginning of every year, December will always start off with 6 days only.

The year

The year will be 360 days long or 12 months of 30 days each or 51 weeks and three days.

Appendix A

The idea of the pi-cycle developed when I computed the ratio of the distance between the earth and sun to the distance covered by the earth in its orbit around the sun in a single rotation or one day. I found out that this ratio was close to one radian and I got suspicious that something we do not know could be hidden in this ratio. Further refinement of the numbers used in the computation revealed that this ratio is actually equal to one radian.

By definition, the pi-cycle is a ratio. Defined in broad terms of a circle, it is the ratio of multiplying a constant number 180 by the time it takes the particle or object in a circular motion around a point to generate 1 degree with respect to the centre of the circular motion multiplied by the velocity of the particle or object in the circular motion, then divided by its distance from the centre. This ratio is a constant that is equal to pi (π).

The equation below is the mathematical expression of the above statement:

$$\pi = \frac{180vt}{r}$$

Appendix B

Generating the Day

The sun-earth system calculations generate the general equation of time for the earth. The sun-earth system calculations are used to derive the general equation of time with the assumption that the sun-earth system is the fundamental time-generating system. This makes Earth the time base generator for the solar system and universe at large.

It starts from the definition of the pi-cycle, which is really a simple formula for pi (π) based on a particle tied to a string and in a circular or angular motion around centre o.

The statement for the formula for π states that a particle a at a distance r is in a circular motion around a centre o in direction s, at velocity v units per second, minute, hours or even days; the ratio of the time it takes for the particle to resolve 1 degree with respect to the centre o multiplied by its angular velocity in the direction s multiplied by 180 and divided by the radius r from the centre is equal to a constant π—this is stated mathematically in the following formula:

$$\pi = \frac{180vt}{r} \quad (i)$$

This formula for the pi-cycle is executed every time the earth completes one 360° rotation on its axis and the formula is true for any particle in a circular motion around centre o at any distance and velocity under the following conditions:

1. $\Delta s \neq 0$ or - 1
2. $\Delta v \neq 0$ or - 1
3. $r \neq 0$ or - 1

To validate the formula, a dimensional analysis on the equation is performed as follows:

r is a measure of distance d and therefore can be replaced by d in the denominator.

v is a derivative of distance r divided by time t or d/t; stated simply, it is the distance travelled over a period of time divided by the time it took to travel the distance so we replace v with d/t then simplify the fraction.

Substituting the above quantities in the equation for π, we get the following fraction:

$$\pi = \left[\frac{180}{\frac{d}{t}} \times \frac{t}{1}\right] / d \quad (ii)$$

Simplifying the fraction, we get:

$$\pi = 180 \quad (iii)$$

Then from the equation of π:

$$t = \frac{\pi r}{180} \quad (iv)$$

And:

$$v = \frac{\pi r}{180 t} \quad (v)$$

The other equation of velocity of interest is the equation of the velocity of an artificial or natural satellite around a centre of gravity:

$$v = \sqrt[2]{\frac{Gm}{R}} \quad (vi)$$

Since the two equations, $v = \sqrt{(Gm/R)}$ and $v = \pi r/180t$, are used to calculate the quantity velocity, then they are equivalent to each other if the resulting velocity calculations are the same or within reasonable error limits. To create the general equation of time, the centre o is replaced with a centre of gravity and the two equations:

$$v = \frac{\pi r}{180t} \quad (v)$$

and:

$$V = \sqrt[2]{\frac{Gm}{R}} \quad (vi)$$

are combined to solve for t.

Solving for t gives us the equation for the time it takes the planet earth to resolve 1° with respect to the centre of the sun and it turned out to be the time it takes the planet earth to rotate 360° on its axis and equivalent of one 24-hour day or the pi-cycle.

A 360° rotation of the earth on its axis is exactly 1 day; this means that for the first time in the history of time measurement, there is an equation to calculate the length of day as opposed to the current method of estimating it by observing when the "upper" limb of the sun appears above the horizon, or using a fictitious star.

The next step is to merge the two equations (i) and (iv), the standard equation for computing the velocity of an artificial or natural satellite in motion around a centre of gravity, like the earth around the sun or the moon around the earth, and find time t in terms of the other variables. The reason for this has been mentioned earlier as being that equation (v) for computing velocity derived from the equation (i) for computing π is equivalent to equation (vi) because both equations are calculating the same quantity, velocity.

We start by computing the velocity of the planet in its orbit around the sun using the equation:

$$V = \sqrt[2]{\frac{Gm}{R}}$$ at a distance of 149,597,870,700 metres between the earth and sun.

We have:

R = 149,597,870,700 metres
Mass of the sun M = 1.9×10^{30} kg
Gravitational Constant G = 6.67×10^{-11}

$$V = \left(\sqrt{\dfrac{\left(6.67 \times 10^{-11} \text{ N}\dfrac{m^2}{kg^2}\right) \times (1.9 \times 10^{30} \text{ kg})}{149{,}597{,}870{,}700}}\right)/3600$$

$V = (\sqrt[2]{850919463.0872483221476510067141})/3600$ metres per second

$\quad = 29122.595341682$ metres/second
$\quad = 29.122595342$ km/second
$\quad = 104841.343230055$ km/hour
$\quad = 65525.83951878$ miles/hour

However, using the velocity obtained using the equation and $V = \sqrt[2]{\dfrac{Gm}{R}}$, it is not possible prove that 1 loop around the sun by the planet earth is equal to 360 degrees in a circle or an ellipse. The solution to this problem is to merge equation $v = \sqrt[2]{\dfrac{Gm}{R}}$ and $v = \dfrac{\pi r}{180t}$ and solve for time t, resulting in the general equation of time.

The general equation of time

In this section, we will merge the two equations, $v = \sqrt[2]{\dfrac{Gm}{R}}$ and $v = \dfrac{\pi r}{180t}$, to solve for time t and the resulting equation will predict the amount of time it takes for the planet earth to rotate on its axis to generate a day as we know it. Before this new equation, there has been no equation for calculating the length of the day that is equivalent to the sidereal day.

Merging the two equations:

$$v = \sqrt[2]{\dfrac{Gm}{R}} \text{ and } v = \dfrac{\pi r}{180t}$$

To merge the two equations, we write them side-by-side and divide each side by V to eliminate it from both equations and equate them to solve for t.

Therefore:

$$v = \sqrt[2]{\frac{Gm}{R}} \quad \text{and} \quad v = \frac{\pi r}{180t}$$

Divide both sides by **V** to equate them and then solve for t:

$$\frac{\pi r}{180t} = \sqrt[2]{\frac{Gm}{R}}$$

Square both sides to get rid of the square root:

$$\frac{r^2 \pi^2}{180^2 t^2} = \frac{Gm}{R}$$

Cross-multiply and equate both sides:

$$R\pi^2 r^2 = 180^2 t^2 Gm$$

Solve for t:

$$t^2 = \frac{\pi^2 R^3}{Gm \, 180^2}$$

Divide π^2 by 180^2 to get rid of the fraction, as π^2 and 180^2 are constants that never change, to create a constant **k**.

Therefore:

$$k = \frac{\pi^2}{180^2} = \frac{(3.14159265358979323846264338327 95)^2}{180^2}$$

Therefore:

$$k = \frac{9.869604401089358618834490999 8762}{32400}$$

$$= 3.0461741978670859934674354937889 \times 10^{-4}$$

Therefore:

$$t^2 = \frac{kR^3}{Gm}$$

Take the square root of both sides to solve for t:

$$t = \sqrt{\frac{kR^3}{Gm}}$$

Therefore, the general equation of time in seconds is defined as:

$$t = \sqrt{\frac{kR^3}{Gm}}$$

And finally, for the general equation of time, $t = \left(\sqrt{\frac{kR^3}{Gm}}\right)/3600$ seconds.

In order to calculate the length of the day using the general equation of time, we need to evaluate **t** using the distance between the sun and earth measured centre to centre from earth to sun or simply, the absolute surface-to-surface distance between the earth and sun. We will use both distances for comparison:

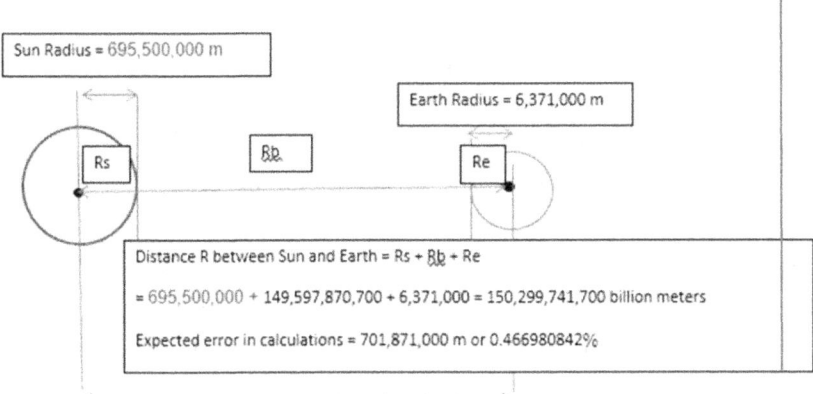

Calculating the length of the day using centre-to-centre distance between the earth and sun

Effective distance between the sun centre and earth centre = 150,299,741,700 billion metres:

Where:

$k = 3.046174198 \times 10^{-4}$ *The value of k coming from* $\left[\dfrac{\pi^2}{180^2}\right]$ And:

R = 150,299,741,700 billion metres
M = 1.989 x 10^{30}
G = 6.67 x 10^{-11}

Substituting the above numbers in the equation, $t = \left(\sqrt{\dfrac{kR^3}{Gm}}\right)/3600$ seconds, we have:

$$\left(\sqrt{\dfrac{3.046174198^{-4} \times 1.50299741700 \times 10^{11} \times 1.50299741700 \times 10^{11} \times 1.50299741700 \times 10^{11}}{6.67 \times 10^{-11} \times 1.989 \times 10^{30}}}\right)/3600$$

$$= \left(\sqrt{\dfrac{1.034259307 \times 10^{30}}{1.326663000 \times 10^{20}}}\right)/3600$$

$$= (\sqrt{7795945971})/3600$$

= 88,294.65426/3600

= 24.52659285 hours

Adjust for the error of 0.466980842%:

= 24.52629285 - (24.5262985 x (0.466980842/100) = .114533079

= 24.41175977 hours or 24 hours, 24 minutes and 42 seconds

Compared to the current length of the sidereal day of 23 hours, 56 minutes and 4 seconds, we have an error of 28 minutes and 38 seconds.

If we use the effective distance of 149,597,870,700, **t** will evaluate to 24.35469395.

Effective distance between the sun centre and earth centre = 149,597,870,700 billion metres, where:

$k = 3.046174198 \times 10^{-4}$ *The value of k coming from* $\left[\frac{\pi^2}{180^2}\right]$ And:

$R = 149{,}597{,}870{,}700$ billion metres
$M = 1.989 \times 10^{30}$
$G = 6.67 \times 10^{-11}$

Substituting the above numbers in the equation, $t = \left(\sqrt{\frac{kR^3}{Gm}}\right)/3600$ seconds, we have:

$$\left(\sqrt{\frac{3.046174198^{-4} \times 1.49597870700 \times 10^{11} \times 1.49597870700 \times 10^{11} \times 1.49597870700 \times 10^{11}}{6.67 \times 10^{-11} \times 1.989 \times 10^{30}}}\right)/3600$$

$$= \left(\sqrt{\frac{1.019837486 \times 10^{30}}{1.326663000 \times 10^{20}}}\right)/3600$$

$$= \left(\sqrt{7687238479.441478073}\right)/3600$$

$= 87676.898208373/3600$

$= 24.3546939467$ hours

$= 24.41175977$ hours or 24 hours, 21 minutes and 17 seconds

Compared with the current length of the sidereal day of 23 hours, 56 minutes and 4 seconds, we have an error of 25 minutes and 13 seconds.

At this point, we can use this **t** to test the formula for pi again: $\pi = \frac{180vt}{r}$

We have $v = \frac{\pi r}{180t}$

Because the two equations are equivalent, we can now evaluate v using equation, $v = \frac{\pi r}{180t}$, and the effective distance between the sun and earth of 150,299,741,700 billion metres.

We have:

$$V = \left(\frac{3.141592654 \times 150{,}299{,}741{,}700}{180 \times 24.41175977 \times 60 \times 60}\right)$$

$$V = \left(\frac{4.721805644^{11}}{15{,}818{,}820.33}\right)$$

= 29849.229056 metres/second
= 29.84929056 km/second
= 107457.4460 km/hour
= 67160.90376 miles/hour

If we use the effective distance of 149,597,870,700 metres between the sun and earth, we have:

$$v = \frac{\pi r}{180t}$$

$$V = \frac{3.141592654 \times 149{,}597{,}870{,}700}{180 \times 24.35469395 \times 60 \times 60}$$

$$= \frac{4.699755716 \times 10^{11}}{15{,}781841.68}$$

= 29779.51377 m/sec
= 29.77951377 km/sec
= 107206.2496 km/hour
= 67003.90598 miles/hour

Validating the equation of pi

At this point, let's validate the equation of pi using the above velocity and using centre-to-centre distance between the earth and sun and surface-to-surface distance between the earth and sun.

Start with a centre-to-centre distance of 150,299,741,700 metres between the earth and sun:

$$\text{Using the formula } \pi = \frac{180vt}{r}$$

Let us compute pi using:

$$v @ 150{,}299{,}741{,}700 = 29779.51377 \text{ meters per second}$$
$$t @ 150{,}299{,}741{,}700 = 87882.335172 \text{ seconds}$$

$$\pi = \frac{180 \times 29849.229056 \times 87882.335172}{150299741700}$$

$$= \frac{4721795914545.3476837376}{150{,}299{,}741{,}700} = 3.141586180$$

$$= 3.141586180$$

Then let us validate the equation for pi using the surface-to-surface distance of 149,597,870,700 metres between the earth and sun. Let us compute pi using:

$$v @ 149597870700 = 29779.51377 \text{ meters per second}$$
$$t @ 149597870700 = 87676.89820812 \text{ seconds}$$

$$\text{Using the formula } \pi = \frac{180vt}{r}$$

$$\text{We have } \pi = \frac{180 \times 29779.51377\ 7 \times 87676.89820812}{149{,}597{,}870{,}700}$$

$$= \frac{469{,}975{,}571{,}549.927615846232}{149{,}597{,}870{,}700} = 3.141592653$$

The orbital velocity computed using the equation $v = \sqrt[2]{\frac{Gm}{R}}$ at a distance of 149,597,870,700 metres between the earth and sun is lower by 1,770 miles/hour or 2,360 km/hour compared to the velocity computed by the equation:

$$\pi = \frac{180vt}{r}$$

Which is derived from the equation of pi (π) and is the real reason why astronomers have been unable to close the planet's loop around the sun to equal 360°.

Validating the Equation of Π Using Two Different Velocities

At this point, lets validate the equation of pi using the above velocity and centre-to-centre distance between the earth and sun, then surface-to-surface distance between the earth and sun.

Starting with the centre-to-centre distance of 150,299,741,700 metres between the earth and sun, the following calculation will validate the formula for pi:

$$\pi = \frac{180vt}{r}$$

Pi (π) can be computed using the following values of v and t:

v @ 150,299,741,700 = 29779.51377 m/s
t @ 150,299,741,700 = 87882.335172 seconds

We have:

π = (180x 29849.229056x 87882.335172)/150,299,741,700
= (4721795914545.3476837376)/150,299,741,700
=3.141586180

Then let us validate the equation for pi using the surface-to-surface distance of 149,597,870,700 metres between the earth and sun using the formula, π = 180vt/r or the pi-cycle.

Pi (π) can again be computed using the following values of v and t:

v @ 149,597,870,700 = 29779.51377 metres per second
t @ 149,597,870,700 = 87676.89820812 seconds

We have:

π = (180x 29779.51377x87676.89820812)/149,597,870,700
= 469,975,571,549.927615846232/149,597,870,700
= 3.141592653

The velocity of Earth calculated from both equations is close within reasonable limits but not the same. The only problem is that with the results of 65525.839518786 miles per hour for the velocity of the planet, using the equation, $V = \sqrt{\frac{Gm}{R}}$, at a distance of 149,597,870,700 metres between the earth and sun, astronomers haven't been able to close the 360° loop around the sun.

Evaluating the above equation using the values given with respect to the earth and sun, where the sun is the central gravitational object modulating the planet earth in the orbit and rotating it on its axis at a variable speed and directly proportional to its distance from the sun. The axial speed of Earth on its axis varies almost every second with an average:

$$t = 24.354694946 \text{ hours}$$

This is the time it takes for Earth to generate an angle of one degree with respect to the centre of the sun and it turns out to be the average length of the day.

This value is close, if not equal, to the length of the day depending on the location of the planet in its orbit around the sun. The equation seems to suggest that the planet will spin a little faster when it is closest to the sun and a little slower when it is further away from the sun.

The German astronomer Kepler discovered that the velocity of the planet increases when it is closest to the sun and drops when it is furthest away from the sun. The equation of time simply predicts the time it takes for Earth to rotate 360° degrees on its axis—this is also the first time we get to know this information and as we shall find out later, it plays an important role in predicting the phases of the moon against the phases of the sun.

Further investigation of the equation shows that the period T of the planet minus the time T it takes for the planet to circle the sun can be computed as T = 360t in hours. This means that the period of the earth cannot exceed 360 days from a mathematical point of view; therefore, the length of the year cannot exceed 360 days.

The period of the planet would be 360t:

$$= 360 \times 24.354693946$$
$$= 8767.6898201 \text{ hours}$$

The length of the week would increase to 7.5 days because the sun does not shine on both sides of the planet at the same time. There is a 180 lag in the midday and midnight points on any two locations on the planet that are opposite to each other.

The number of weeks in a year would drop to 48 weeks of 7.5 days each.

Appendix C

The Moon-Earth System

Calculating t for the moon using a formula derived from the pi-cycle

In order to calculate the moon cycle or the length of the month, it is necessary to assume that the moon-earth system is a different system that is a subset of the sun-earth system and the general equation of time can be used to calculate the time it takes for the moon to generate 1° with respect to the centre of the earth.

The reason for this is because the number of days in a month is a subset of the days generated by the earth when it generates 1° with respect to the centre of the sun.

Therefore, for the moon:

$$t = \left(\sqrt{\frac{kR^3}{Gm}}\right)/3600$$

Where:

Me = mass of the earth = 5.972×10^{24}
G = the gravitational constant = $6.666666667 \times 10^{-11}$
And $k = 3.046174198 \times 10^{-4}$

$R = r_1 + r_2$
$r_1 = 3.845 \times 10^8$ (Distance from earth to moon in metres)
$r_2 = 6.371 \times 10^6$ (Radius of the earth in metres)
$R = 3.845 \times 10^8 + 6.371 \times 10^6 = 3.90\,871 \times 10^8$

Then t for the moon in seconds = $t = \left(\sqrt{\dfrac{kR^3}{Gm}}\right)/3600$

$= \left(\sqrt{\dfrac{3.046174198 \times 10^{-4} \times 3.90871 \times 10^8 \times 3.90871 \times 10^8 \times 3.90871 \times 10^8}{6.67 \times 10^{-11} \times 5.972 \times 10^{24}}}\right)/3600$

$= \left(\sqrt{\dfrac{1.8190938 \times 10^{22}}{3.98133335324 \times 10^{14}}}\right)/3600$

$= \left(\sqrt{45690566.498064178}\right)/3600$

$= 6759.47976543/3600$
$= 1.87763326815$ hours

Therefore, **tmoon** or the time it takes the moon to resolve 1° with respect to the centre of the earth is = 1.87763326815 hours.

Now that t for the moon is known, v for the moon in its orbit around the earth can be calculated using the formula:

$$v = \dfrac{\pi r}{180t}$$

$$v = \dfrac{3.141592654 \times 3.90871 \times 10^8}{180 \times 6759.47976543}$$

$= 1009.247182$ metres/second
$= 3633.289854$ km/hour
$= 2270.806158$ miles/hour

After computing the orbital velocity of the moon around the earth, π can be computed from the formula:

$$\pi = \dfrac{180vt}{r}$$

$$\pi = \dfrac{180 \times 009.247182 \times 6759.47976543}{3.90871 \times 10^8}$$

$$= \frac{1227957462.9}{3.90871 \times 10^8}$$

$$= 3.1415926556$$

Next, the moon/earth cycles for the 365.25 and 360-day years are calculated using both the sidereal and calculated lengths of the day. The following calculations use the sidereal day.

Now T, the period or moon cycle or the time it takes the moon to circle the earth, can be calculated from the formula:

$T = 360(t_{moon})$ divided by 23 hours, 56 minutes, 04 seconds—the official length of the day.

Therefore, T for the moon - the time it takes the moon to circle the earth:
$T = 360(t_{moon}) = 360 \times 1.87763326815 = 675.94797653$ hours

T in days = 675.94797653/23.93447139 or (23 hours, 56 minutes, 04 seconds—the official length of the day)
Then $T = 28.241608745$ days.

If we multiply the calculated number of days it takes for the moon to circle the earth and multiply it with number of months in a year, we shall get fewer days for both the 365.25 and 360-day years.

For the 365.25-day year

If the length of months was 28.241608745 days, then $28.241608745 \times 12 = 340.135237031$ would equal to days in a year.

So for the 365.25-day year, we are short 25.114763 days or 25.114763/12 = 2.09289692 days missing per month.

If the 2.09289692 missing days per month are added to the 28.344603086 moon cycle, the result is 30.4375 days in a moon cycle or month for the 365.25-day year.

By observation, there is one missing day in every moon cycle because the moon loses an average of 48 minutes for every 360-degree rotation of the earth on its axis. This works out to be:

48 minutes x 31 days for the month of January = (48x31)/(23.93447139x60) = 1.036162428 missing days for every 31-day month = (1.036162428 x 7) + (1.036162428 x 4) +
(48x28)/(23.93447139x60) for non-leap years = 7.253136996 + 4.144649712 + 0.935888645 =
12.33552273 missing days for non-leap year
or = (1.036162428 x 7) + (1.036162428 x 4) + (48x29)/(23.93447139x60) for the leap years
= 7.253136996 + 4.144649712 + 0.969313239 = 12.36894732 missing days in a leap year

Therefore, for the 365.25-day year calendar, we need to find an additional 25.114763 - 12.33552273 = 12.77924027 missing day after accounting for the monthly loss for the non-leap year. 25.114763 - 12.36894732 = 12.74581568 additional missing day after accounting for the monthly loss for the leap year. The above two calculations show the missing days in a leap year or non-leap year.

For the 360-day year

For the 360-day year, we are short 19.8647629692 days or 19.864762692/12 = 1.65539694 days missing per month.

If the 1.65539694 missing days per month in a 360-day year are added to the 28.344603086 moon cycle, we get 30.0000000027 days in a moon cycle or month.

Calculating the moon/month cycle using the calculated day length
For the 365.25-day year
Therefore, T for the moon is:

$T = 360(t_{moon}) = 360 \times 1.87763326815 = 675.94797653$ hours

T in days = 675.94797653/24.354693947 hours or (24 hours, 21 minutes and 17 seconds, the calculated length of the day) = 27.855053758 days.

Then T = 27.855053758 days.

If we multiply the calculated number of days it takes for the moon to circle the earth and multiply it with number of months in a year, we shall get fewer days than the 365.25 days.

So if the month was 27.754320296 days x 12 = 333.05184355 days in a year.

For the 365.25-day year, we are short 365.25 - 333.05184355 = 32.19815645 missing days in a year or 2.683179704 missing days per month.

Now if we add the 2.683179704 missing days per month to the 27.855053758 calculated moon cycle, we get 30.53823362 days in a moon cycle or month.

By observation, there is only one missing day in every moon cycle for a total of 12 days.

The other 18.989355 days a year are missing or 1.582444625 days missing from each month are hidden and they must be found by observation and measurement in order to balance the 365.25-day year with the 30.4375 days moon cycle or month.

For the 360-day year

Therefore, T for the moon is:

$T = 360(t_{moon}) = 360 \times 1.87763326815 = 675.94797653$ hours

T in days = 675.94797653/24.354693947 hours or (24 hours, 21 minutes and 17 seconds, the calculated length of the day) = 27.754320246 days

Then T = 27.754320246 days.

If we multiply the calculated number of days it takes for the moon to circle the earth and multiply it with number of months in a year, we will get fewer days than the 360 days.

If the month was 27.754320296 days long, then 27.754320296 x 12 = 333.05184355 days in a year.

For the 360-day year, we are short 360 - 333.05184355 = 26.94815645 days or 2.245679704 days missing per month.

Now if we add the 2.245679704 missing days per month to the 27.754320296 calculated moon cycle, we get 29.99999995 (30) days in a moon cycle or month.

By observation, there is only one missing day in every moon cycle for a total of 12 days.

The other 13.7393549047 days a year missing or 1.144946242 days missing from each month are hidden and they must be found in order to balance the 360-day year with the 30-day moon cycle or month.

Variations in the moon's orbital velocity

The moon-tracking instrument will record the progress of the moon in its orbit using successive additions of the angle generated by the moon with respect to the centre of the earth for every 360-degree rotation on the earth on its axis.

It can be seen from this data that the moon advances an average 12 degrees for every one 360-degree rotation of the earth on its axis. However, one has to keep in mind that orbit of the moon around the earth is not a perfect circle; therefore, to validate this data, we need to design and deploy an instrument that will track the moon one rotation of the earth at a time for a minimum of two years to generate data that can be used to automate the tracking of the moon in its automatically.

The moon-tracking instrument will detect the variations in the moon velocity for earth's every rotation and careful analysis of the data should detect when the moon is in the perihelion—the point in the orbit of the moon when it is nearest to the earth—and aphelion—when it is the furthest from the earth.

With a little luck and the resulting new mathematical analysis of the data, it will be possible to detect the nature of its oscillating orbit around the earth.

It is this oscillating nature of the earth and moon orbits that caused the three-day delay in the start of Ramadan fasting in June 2015 because both the moon and earth were in the perihelion with respect to their orbital foci.

Appendix D

The moon demystified

The belief that the moon cycle is 29.5 days long is absolutely correct regardless of the calculations below. Using the angular tracking of the moon, we will find that at some point in its orbit around the earth, the moon will appear be in a 180-degree alignment with a point in space. When this happens, the moon will be between the earth and the sun and will be larger than average, except the super moon. An observer on earth can track the moon throughout the night until it sets in the west early in the morning. In other words, the moon will be sinking below the horizon.

As with Sirius, the next major phase before the moon completes its loop around the earth will be to rise with the sun from the same direction to the observer—east at an angle I am estimating to be about 330 degrees or more but less than 360 degrees. When this happens, the moon will cross the sky in phase with the sun or, for a better explanation, the moon will be hidden in the glare of the sun. According to the Egyptians, Sirius' heliacal risings followed by Sirius disappearing for about 70 days after which it re-appears in the west are referred to as Sirius heliacal settings. Before the moon rises with the sun, its waning crescent will be facing the upcoming morning sun and during one of the earth's rotations, the moon will be positioned exactly between the earth and the sun for .5 of a day. After this phenomenon, the next phase of the moon will be a very small crescent facing the sun. Until this observation is accepted by the scientific—especially the astronomical—community, I will call this the heliacal setting of the moon. We don't see the heliacal rising of the moon because it is hidden in the sun's glare for 12 hours or 0.5 day. If 0.5 of a day is added to the 29.5 days, we get the 30 days of the moon cycle without calculations or measurements.

The only difference between the above explanation and the current illustrations of the phases of the moon is that the former displays the moon phases as static and keeps the sun in the same direction throughout the entire moon cycle. That is positively not true because the moon and earth are two dynamical systems orbiting the sun as a team. The moon probably stays in conjunction with the sun for a minimum of 12 hours and we pass it in the evening as a new moon because the current tracking methods of the moon are: first sighting to the next sighting and with respect to a fixed star. Astronomers have been omitting the conjunction phase by misinterpreting the static tracing illustration in books of astronomy.

And finally, if my explanation for the 0.5 being hidden in the sun's glare is correct, it should not be difficult for optical engineers to design a special-purpose moon telescope that will see the moon dynamically located into the sun by the rotation of the earth and register that missing half a day.

Appendix E

Moon tracking table, 10 December 2016 to 11 January 2017, showing the 30 days of the moon cycle:

This preliminary naked-eye angular tracking of the moon in its orbit around the earth shows that after 30 rotations of the earth on its axis, the moon generates 360 degrees in its orbit around the earth.

On 9 January 2017, the moon was clearly visible in the same location as it was observed on 10 December 2016. The reason the moon was in the same location is because the counting of the earth's rotations starts from 0 and the angle generated by the moon is also 0.

Earth's rotation	Angle generated by the moon	Date	Comments
0	0	10/12/2016	New moon
1	12	11/12/2016	
2	24	12/12/2016	
3	36	13/12/2016	
4	48	14/12/2016	
5	60	15/12/2016	First quarter
6	72	16/12/2016	
7	84	17/12/2016	
8	96	18/12/2016	
9	108	19/12/2016	
10	120	20/12/2016	Second quarter
11	132	21/12/2016	
12	144	22/12/2016	
13	156	23/12/2016	

14	168	24/12/2016	
15	180	25/12/2016	Full moon
16	192	26/12/2016	
17	204	27/12/2016	
18	216	28/12/2016	
19	228	29/12/2016	
20	240	30/12/2016	Third quarter
21	252	31/12/2016	
22	264	01/01/2017	
23	276	02/01/2017	
24	288	03/01/2017	
25	300	04/01/2017	Last quarter
26	312	05/01/2017	
27	324	06/01/2017	
28	336	07/01/2017	
29	348	08/01/2017	Moon rises with the sun?
30	360	09/01/2017	New moon cycle starts

Appendix F

Heliacal risings and settings of Sirius demystified

According to Lynn E. Rose, in *Sun, Moon and Sothis*, after the heliacal rising of Sirius in Ancient Egypt, it went out of sight for about 70 days every year. This is partially true or should be referred to as the "apparent disappearance of Sirius" because in reality, the star is still in its designated location but it has been masked by the sun's glow due to increasing area illuminated by the sun in the northern hemisphere during summer.

The best interpretation of this statement should be "during the 70 days or so, the planet pulls up Sirius and the star is masked by the glare of the sun as the area illuminated by the sun in the northern hemisphere increases as the planet heads towards northern summer in its orbit around the sun". This is because when the planet approaches the summer region of the northern hemisphere, the earth surface illuminated by the sun increases and as a result, to the observer on earth, the sun rises earlier than Sirius. However, the fact is that the planet is getting further away from Sirius while at the same time, the star is hidden in the sun's glare. The area illuminated by the sun keeps on increasing and reaches its maximum on the summer solstice on 21-22 June and will appear to stay stationary until the earth changes direction to start heading for the winter solstice around 15 July, but this will vary over several years as the oscillating loop of the earth's orbit around the sun changes.

When the planet starts to head for winter in the northern hemisphere, the area illuminated by the sun in the northern hemisphere starts to pull back while the planet is actually passing the star during the day when it is hidden in the illuminated area of Earth. Then the observer will see the heliacal setting of the Sirius after around 70 days, more or less, again depending on the location of the planet in the oscillating loop around the sun, when the illumination of the planet starts pulling back as the planet heads into northern winter.

After the heliacal settings of Sirius, the observer will be waiting for heliacal risings again; however, more precise observations will be needed to plot the location of Sirius mile by mile or kilometre by kilometre after the heliacal settings or risings. This explanation can be simplified by stating that the heliacal rising and setting of Sirius is nothing more than the regular pushing up and down of the area illuminated by the sun on the planet to create summer in the northern hemisphere and pulling back the same area to create winter in the northern hemisphere. The reverse is true for the southern hemisphere. As the area illuminated by the sun increases to create summer in northern hemisphere, the area illuminated by the sun is pulling back in the southern hemisphere to create southern winter and vice versa to create southern summer. This explains why Bob King's table in *Sky and Telescope* shows the heliacal rising of Sirius appearing later and later from Boston in summer, maxing out on 21 August at 50 degrees north.

Appendix G

The 365.25 to 360-Day Year Calculator

This is a 365.25-day year to 360-day year conversion calculator.

The 365.25-day year to 360-day year converter calculator takes the 365.25-day year and converts it into a 360-day year. For now, it works with only whole number years starting from the end of year 118 or 119 BCE to the end of year 5,000 BCE and works with positive integers only because time is infinitely incremental and the past cannot be predicted because it has already occurred.

For now, the converter uses a 24-hour clock only when converting hours, minutes and seconds.

The 365.25-day year to 360-day year converter takes advantage of the fact that when the Roman Emperor, Julius Caesar, reformed the Roman calendar in 45 BCE by adopting the Egyptian calendar that had been reformed 5 years earlier in 50 BCE by the Egyptian Moon God Thoth, Caesar added 5.25 days to the Roman calendar and sprinkled them in the months of January, March, May, July and August by giving these months 31 days each with .25 a day accumulating every year, creating a leap year every 4 years. Caesar's calendar fell apart by year 1582 and the church granted Pope Gregory the honour of fixing it. What Pope Gregory did to fix the calendar at that time appears to be the same as what Julius Caesar had done in 45 BCE, meaning that the 5.25 extra days in the calendar remained untouched.

The 365.25-day year to 360-day year converter will covert 2015, a year with 365.25 days, and convert it into a 360-day year and the resulting 360-day year will be: 2044, 4:18:18:0:0—year 2044, March 18th at 18 hours, 00 minutes, 00 seconds.

If a user wants to convert a series of years by entering the start year and end year—for example, from 1900 to 2000—it will display a series of years converted linearly from 1900 to 1970. Noticeable in this conversion is the convergence of Caesar's virtual year with the 360-day year occurring when 365.25-day year 1920 is converted to the 360-day year, giving the result of 1948, 0: 0: 0: 0, meaning that year 1920 in the 365.25-day calendar was actually year 1948, January 1st at 00 hours, 00 minutes and 00 seconds. (Please note that year 0 does exist despite the fact that it was skipped when moving from BCE to A.C.E because it is possible that at that time, 0 was not considered to be a number.)

The next year of the 360-day year, immediately after 1948, 0:0:0:0, starts off 5.25 days ahead of the 365.25-day year at 1949, 0:5:6:0:0 seconds rather than 1949, 0:0:0:0, meaning:

Year 1949, January, 5th at 06 hours, 00 minutes and 00 seconds.

Showing the 5.25 days as extra days that continue to accumulate for 480 years to the next convergence of Caesar's virtual planet and the planet we live on.

Appendix H

Proof that one revolution in any circular motions like circles and ellipses and around any shape, which may not be a perfect circle or ellipse, is equal to 360:

$v = \dfrac{\pi r}{180t}$ is equivalent to $v = \omega r$

$$v = \dfrac{\pi r}{180t} = v = \omega r$$

Divide both sides by v to eliminate it:

$$\dfrac{v}{v} = \dfrac{\pi r}{180t} = \dfrac{v}{v} = \omega r$$

$$\dfrac{\pi r}{180t} = \omega r$$

$$= \dfrac{\pi r}{180t} = 2\pi f r$$

Replace ωr with $2\pi \dfrac{1}{t} r = \dfrac{2\pi r}{t}$

Divide both sides by r to eliminate it:

$$\dfrac{1}{r} \times \dfrac{\pi r}{180t} = \dfrac{1}{r} x = \dfrac{2\pi r}{t}$$

Multiply both sides by 180t to eliminate it:

$$= 180t \times \dfrac{\pi}{180t} = \dfrac{2\pi \times 180t}{t}$$

$\pi = 360\pi$

Divide both sides with π $\quad = \dfrac{\pi}{\pi} = 360$.

Therefore, $1 = 360$.

This simply means that one loop around a circle = 360 degrees.

Appendix I

Distance of the earth from the sun

Four hundred million years ago, the time it took for the planet to spin on its axis was 22 hours. Using the above information, it is possible to calculate how far Earth was from the sun.

Let r_1 = to the distance Earth was from the sun, r_2 be the current distance of the planet from the sun, t_1 be the time it took the planet to spin on its axis, then t_2 is the current time it takes the planet to spin on its axis.

Therefore, t_1 = 22 hours, t_2 = 24.354693946, r_2 = 149,597,870,700 metres.

$$\frac{r_1}{r_2} = \frac{t_1}{t_2}$$

Then: $\quad r_1 = \frac{r_2 \times t_1}{t_2}$

$$= \frac{149,597,870,700 \times 22}{24.354693946}$$

$$= 135,134,240,000 \text{ metres or } 84,458,900 \text{ million miles}$$

As for the length of the year 400 miles years ago, it can be calculated using the same ratios but different numbers.

Let T_1 = length of the year 400 million years ago, T_2 be the current length of the year, R_1 be the distance of Earth from the sun and R_2 be the current distance of the planet from the sun.

Therefore:

87

$$T_1 = \frac{R1 \times T2}{R2}$$

$$= \frac{135134240000 \times 360}{149597870700}$$

$$= 325.19397349 \text{ days}$$

The length of the year in terms of today's time was approximately 325.2 days. However, if we use 24 hours to calculate the distance of the planet from the sun 400 million years ago, we get 329.99997104 days for the length of the year or 330 today's days of exactly 24 hours.

References

1. Hannah, R., (2005) *Greek and Roman Calendars: Construction of Time in the Classical World*, Duckworth: Gerald Duckworth & Co. Ltd.
2. Coveney, P. and Highfield, R., (1990) *The Arrow of Time: A voyage Through Science to Solve Time's Greatest Mystery*, New York: Ballantine Books.
3. Gonzalez, G. P., (2010) *13 B'aktun: Mayan Visions of 2012 and Beyond*, California: North Atlantic Books.
4. Rose, L. E., (1999) *Sun, Moon and Sothis: A Study of Calendars and Calendar Reforms in Ancient Egypt*, Richmond: Kronos Press.
5. Holford-Stevens, L., (2005) *The History of Time*, Oxford University Press.
6. The star Sirius in Ancient Egypt and Babylonia:
7. *http://www.gautschy.ch/~rita/archast/sirius/siriuseng.html*
8. *http://www.skyandtelescope.com/observing/a-real-scorcher-sirius-at-heliacal-rising/*
   This map shows the sky (as seen when facing east) on 15 July 3000 BC from the ancient city of Memphis (near Cairo) in Egypt. Sirius stands 3° high 32 minutes before sunrise around the time of its heliacal rising (latitude 30° 0 N).
   Stellarium

www.ingramcontent.com/pod-product-compliance
Lightning Source LLC
Chambersburg PA
CBHW071418220526
45469CB00004B/1332